Dr. Boo: My First Geometry Tutor

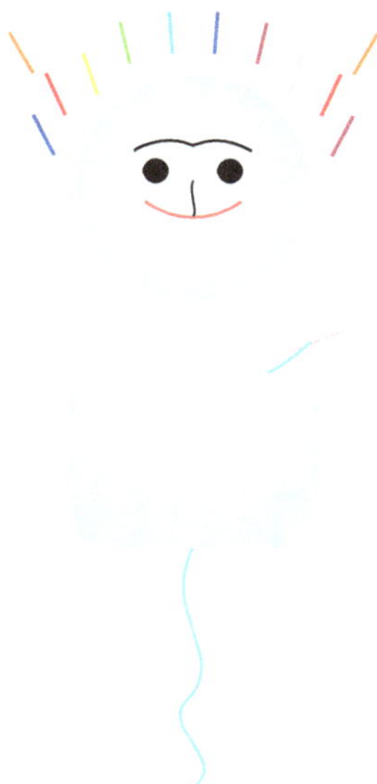

Contents

Chapter 1

Introduction

Every book is a conversation between writer and reader. Thank you for beginning this one with me. I hope the journey ahead feels worthwhile.

Geometry has a reputation for intimidating students long before they ever pick up a compass. It doesn't need to be that way. In fact, geometry is one of the most elegant branches of mathematics—precise, visual, and deeply logical. When it's approached with a bit of curiosity, it stops feeling like a puzzle box and starts feeling like a language anyone can learn.

I wrote this book after watching countless students hesitate over ideas they were absolutely capable of understanding. The problem was never intelligence; it was confidence. A right angle shouldn't feel like a riddle. Parallel lines shouldn't feel like a trick question. These are ideas children can grasp—often faster than adults expect—when they're given room to explore and a clear path to follow. A little imagination doesn't hurt either.

You'll find explanations here, of course, but also diagrams, odd little observations, and the occasional aside that slipped in because I couldn't resist. Some pages move quickly. Others linger. That's intentional. Geometry doesn't unfold in perfect symmetry, and neither does learning.

My hope is that this book helps children meet the subject with less fear and more curiosity. If it gives them the confidence to solve problems on their own—or even to enjoy the quiet logic of a well-drawn shape—then it's done what I meant it to do.

So let's begin. Geometry is far more approachable than it pretends to be, and I'm here to guide your child through it, one idea at a time, until the whole subject feels not mysterious, but familiar.

Dr. Boo

Chapter 2

Points, Lines, and Dad's Steady Guidance

2.1 Points

A point is the simplest thing in geometry. It's just a location. No size, no shape, no drama. Honestly, I wish more things in life were this simple.

Here's a point:

That's it. That's the whole thing. If only your laundry was this easy.

Memory Trick!

A point has **no size at all**. Think of it like the dot on an "i" — tiny, but mighty!

Examples

- Point A is at some location.
- Point B is somewhere else.
- Point C is where Dad steps on a Lego.

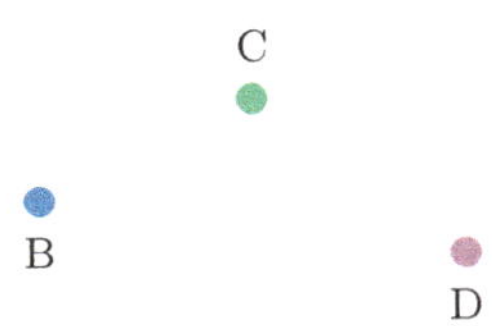

Try It

- Draw three points anywhere on your page.
- Give each point a name.
- Circle the one where you think Dad would step on a Lego.

2.2 Lines

A line goes on forever in both directions. A line goes on forever in both directions — like the number of times I've told my dog to stop chewing my shoes.

Line ℓ

Memory Trick!

A line has **no endpoints**. Imagine it stretching farther than your bedtime excuses.

Examples

- A line never ends. Much like your questions at bedtime.
- Two points determine a line.
- If you put two points anywhere, boom — line.

Line AB

Try It

- Draw two points on your paper.
- Use a ruler to draw the line through them.
- Label the line with a fancy letter, like ℓ or m.

2.3 Line Segments

A line segment has two endpoints. It's like a line, but with boundaries. You know, like the ones you ignore when I say "don't touch the TV remote."

Memory Trick!

A segment has **two endpoints**. Think of them as the "do not cross" cones of geometry.

Examples

- Segment AB has endpoints A and B.
- It's measurable — unlike my patience.

Segment CD

Try It

- Draw a segment and label its endpoints.
- Measure its length with a ruler.
- Draw a shorter segment and a longer segment.

2.4 Rays

A ray starts at one point and goes on forever in one direction. Like when you ask "why?" and it never ends.

Memory Trick!

A ray has **one endpoint**. Picture it like a flashlight beam — it starts somewhere and zooms off forever.

Examples

- Ray AB starts at A and goes through B.
- Rays are half-lines.
- They're great for showing direction — like "go clean your room."

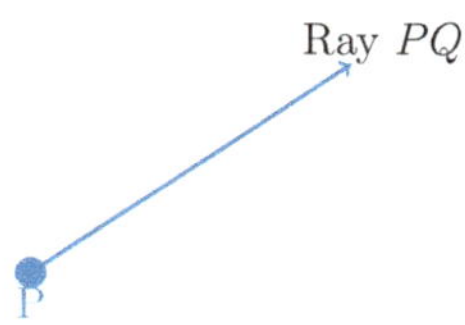

Try It

- Draw a point and name it.
- Draw a ray starting at that point.
- Add arrows to show which way it goes.

Chapter 3

Angles: The Things You Keep Getting Wrong

3.1 What's an Angle?

An angle is formed by two rays with a common endpoint. That endpoint is called the vertex.

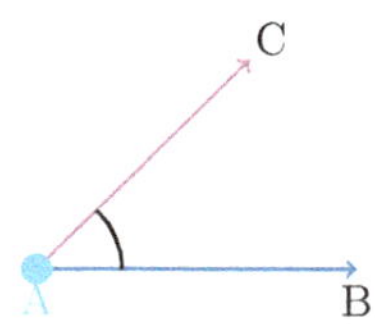

Memory Trick!

An angle is made of **two rays and one vertex**. Think of it like two arms giving a big "V" for victory.

This is angle $\angle BAC$. Or $\angle CAB$. Or $\angle A$.

3.2 Types of Angles

3.2.1 Right Angles

A right angle is exactly $90°$.

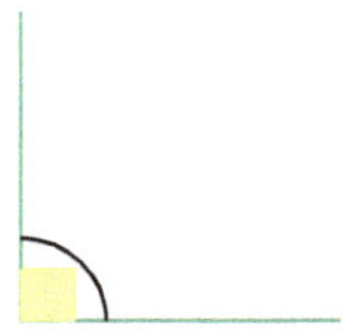

Memory Trick!

Right angles make a perfect **L-shape**. If it looks like a corner you could stub your toe on, it's 90°.

3.2.2 Acute Angles

An acute angle is less than 90°.

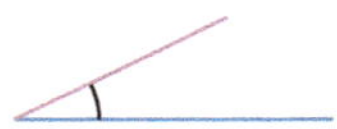

Memory Trick!

Acute = **a-cute**. Small and adorable.

3.2.3 Obtuse Angles

An obtuse angle is more than 90° but less than 180°.

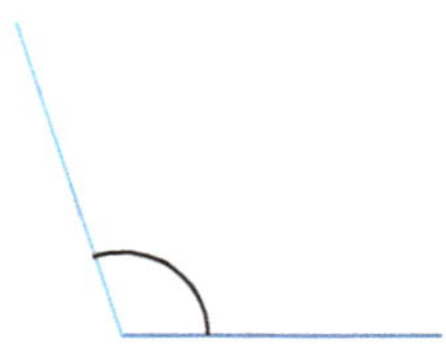

Memory Trick!

Obtuse angles are **bigger than right angles**.

3.2.4 Straight Angles

A straight angle is 180°. It is just a line.

Memory Trick!

A straight angle is just a **line pretending to be an angle**.

Chapter 4

Triangles: Dad's Favorite Shape (Don't Tell the Others)

4.1 What Makes a Triangle?

A triangle has three sides, three angles, and three vertices.

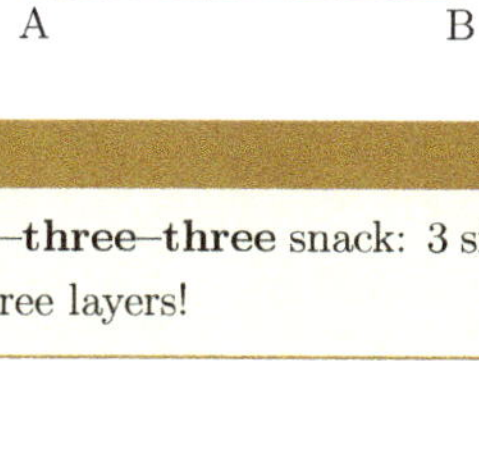

Memory Trick!

Think of a triangle as a **three–three–three** snack: 3 sides, 3 angles, 3 vertices — like a tiny math sandwich with three layers!

Examples

- $\triangle ABC$ has sides AB, BC, and CA.
- Triangles are everywhere — pizza slices, road signs, laundry.

4.2 Types of Triangles

4.2.1 By Sides

Equilateral Triangle

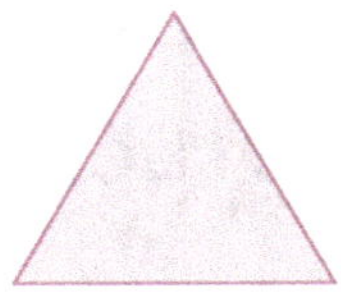

All sides equal

Memory Trick!

Equilateral = Equal! All three sides match perfectly — like triplet siblings who insist on wearing the same outfit.

Isosceles Triangle

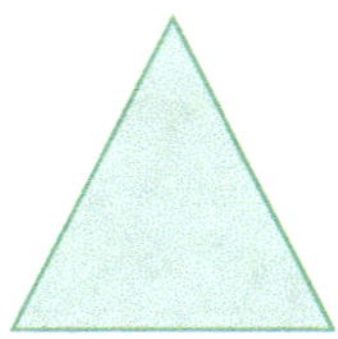

Two equal sides

Memory Trick!

Isosceles has "I–so–same–lees." Say it slowly: "I–so–*same*–lees." Two sides are the same — like two best friends who always walk side-by-side.

Scalene Triangle

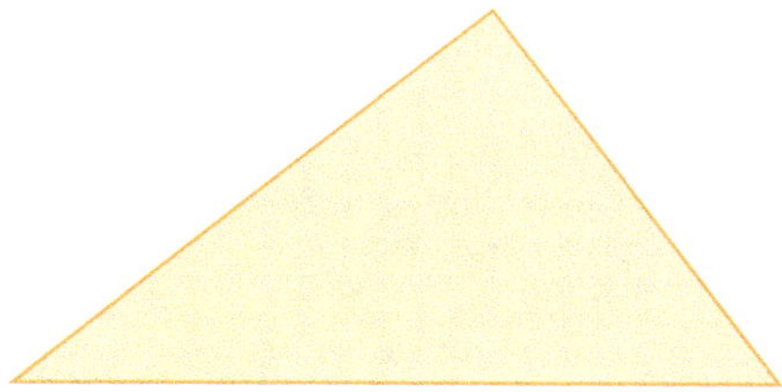

All sides different

Memory Trick!

Scalene = "Scrambled lengths." Every side is a different size — like three crayons that got sharpened at totally different times.

4.2.2 By Angles

Acute Triangle

An acute triangle is one where all three angles are less than 90°. Nothing dramatic here—every corner is sharp, quick, and a little bit shy of a right angle. Students often think "acute" means "cute," and honestly, it kind of fits: these triangles always look a bit small and tidy.

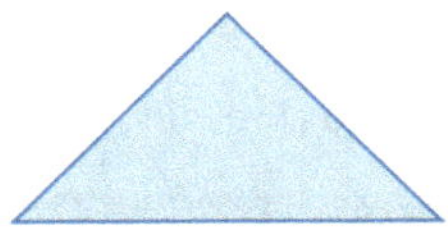

Memory Trick!

Acute = "A-cute little triangle." All the angles are tiny and shy — none of them are brave enough to reach 90°.

Right Triangle

A right triangle has one angle that's exactly 90°. That single right angle changes everything—it gives the triangle a grounded, sturdy feel. You'll see a little square marking the right angle in the figure below, just to make it unmistakable.

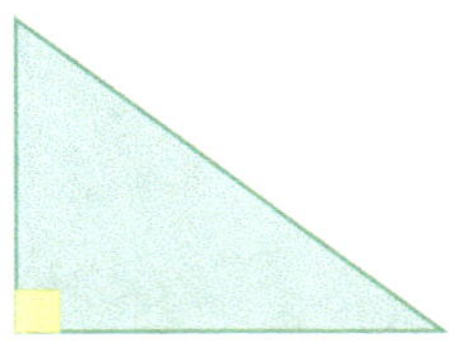

Memory Trick!

Right = "Just right." One angle is a perfect 90° — like the corner of a book or the edge of your desk.

Obtuse Triangle

An obtuse triangle has one angle that's greater than 90°. That single wide angle forces the whole shape to stretch out in a way the other triangles don't. It feels a bit lop-sided—on purpose. Once you spot the "too-wide" angle, you'll never miss it again.

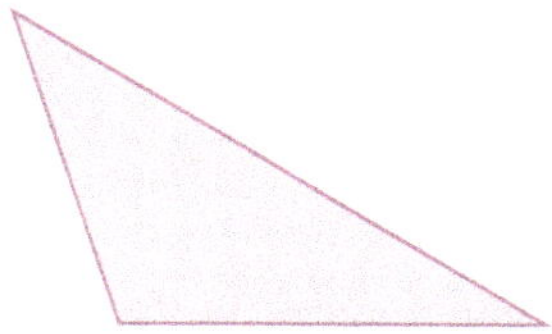

Memory Trick!

Obtuse = "Ob-too-se." One angle is *too wide* — it opens up so far it stretches the whole triangle out of balance.

4.3 Triangle Inequality

4.4 Triangle Inequality

The Triangle Inequality sounds intimidating, but the idea is actually simple: a triangle can only exist if its sides can "reach" each other. If two sides are too short to meet the third, the whole shape collapses into a straight line. No reaching, no triangle.

In math language:

Any two sides of a triangle must add up to more than the third side.

> Memory Trick!
>
> **Memory Trick:** Imagine the three sides as kids trying to hold hands. If two kids can't reach the third, they can't make a triangle "friend circle."

Examples With Diagrams

Example 1: 3, 4, 5 — Works

$$3 + 4 = 7 \quad \text{and} \quad 7 > 5$$

All pairs of sides can reach each other, so this one forms a real triangle.

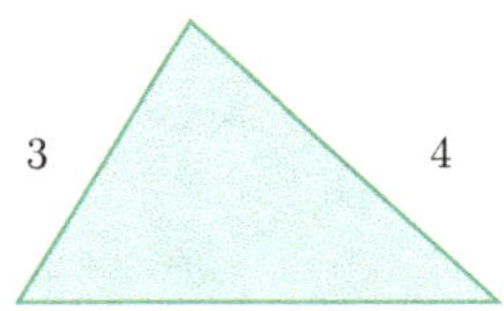

> Memory Trick!
>
> **Why it works:** The two shorter sides (3 and 4) easily reach across the long side (5). They "close the loop," so the triangle stands.

Example 2: 2, 2, 10 — Nope

$$2 + 2 = 4 \quad \text{and} \quad 4 < 10$$

The short sides can't reach the long one. The shape collapses into a line.

Memory Trick!

Why it fails: The two short sides (2 and 2) can't even come close to reaching across the long side (10). No reach = no triangle.

Example 3: 7, 8, 1 — Also Nope

$$1 + 7 = 8 \quad \text{but} \quad 8 \text{ is NOT greater than } 8$$

When the sum equals the third side, the shape flattens into a straight line.

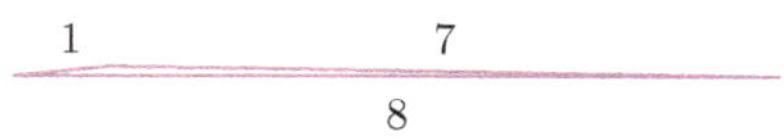

Memory Trick!

Why it fails: The "1" side is too tiny to lift the shape off the line. Everything flattens out, and a flat shape isn't a triangle.

4.5 Dad's Quick Triangle Tips

- **The longest side sits across from the biggest angle.** The triangle "opens" the widest there.
- **The shortest side sits across from the smallest angle.** A tiny angle doesn't open much, so the side across from it stays short.

Memory Trick!

Angle–Side Buddy Rule: Big angle goes with big side. Small angle goes with small side. They always sit across from their matching buddy.

Chapter 5

Congruent Triangles: When Shapes Are Twins

5.1 Ways to Prove Triangles Are Congruent

Before we dive into the alphabet soup of SSS, SAS, ASA, AAS, and RHS, let me say this: congruent triangles are basically shape–twins. Same size, same shape, same everything. If you could pick one up and plop it on top of the other, every side and angle would line up perfectly. It's one of the few moments in geometry where the universe behaves itself.

Now, the shortcuts below are the ways we prove two triangles are truly identical without measuring every last thing. Because nobody has time for that. Not even me.

5.1.1 SSS

If all three sides of one triangle match all three sides of another, the triangles are congruent. No angles needed. No drama. Just three sides doing their job.

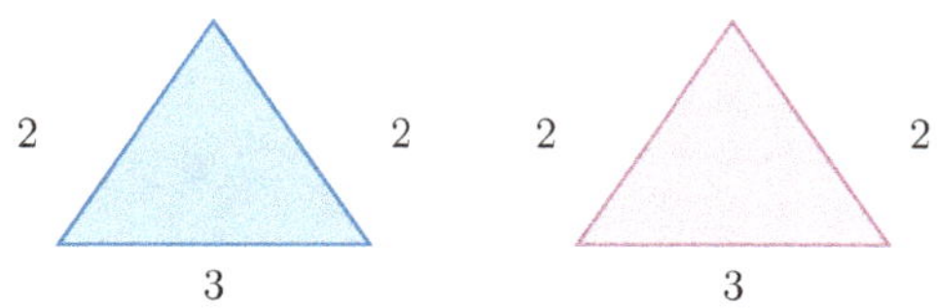

Memory Trick!

SSS Memory Trick: If the "triangle skeletons" match — all three bones — the whole creature matches.

5.1.2 SAS

Two sides and the angle *between* them. Not just any angle — the included angle.

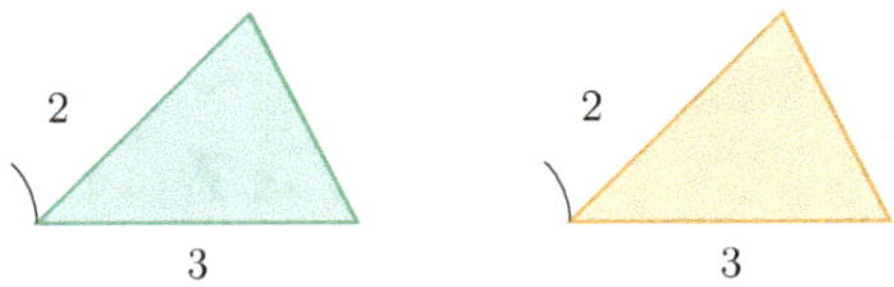

Memory Trick!

SAS Memory Trick: Two sides + the angle between them = triangle locked in place. Miss the included angle and the whole thing falls apart.

5.1.3 ASA

Two angles and the side between them.

Memory Trick!

ASA Memory Trick: Two angles pin the triangle down. The side between them seals the deal.

5.1.4 AAS

Two angles and *any* side.

Memory Trick!

AAS Memory Trick: Two angles already control the shape. Any matching side finishes the job.

5.1.5 RHS

Right angle. Hypotenuse. Side.

Memory Trick!

RHS Memory Trick: Right angle + hypotenuse + one leg = congruent. Right triangles get their own VIP shortcut.

—

Similarity Memory Tricks

AA: "Two angles agree? The third tags along for free."
SSS: "If the sides scale nicely, the triangles play nicely."
SAS: "Two sides in ratio + the angle between = similarity seen."

If the triangle looks like a zoomed-in or zoomed-out version of another, you're probably staring at similar triangles.

—

5.2 Why Similarity Matters

Look, I get it — similarity feels like the quieter cousin of congruence. But it's secretly one of the most powerful tools in geometry. It lets us solve problems we couldn't dream of measuring directly.

- Helps measure things you can't reach. (Want the height of a tree? A building? A dinosaur statue? Similar triangles to the rescue.)
- Used in architecture. (Buildings don't stay upright by accident.)
- Helps solve real-world problems. (Maps, shadows, models, blueprints — similarity is everywhere.)

Chapter 6

Quadrilaterals: Four Sides, Four Times the Chaos

Look, triangles are nice. They're simple. Predictable. Manageable. Quadrilaterals? They're like having four kids instead of three. Suddenly everything is louder, messier, and someone is always upside down.

But we're going to get through this together. Deep breaths.

6.1 What's a Quadrilateral?

A quadrilateral is any shape with four sides. That's it. Four sides. Four angles. Four vertices. Four opportunities for me to question my life choices.

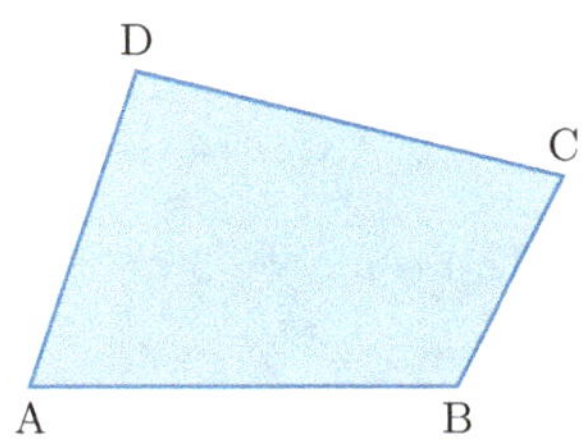

Memory Trick!

Quadrilateral = **quad** (four) + **lateral** (sides). That's literally it. Four sides and a whole lot of personality.

6.2 Types of Quadrilaterals

6.2.1 Parallelogram

Opposite sides are parallel and equal. It's the "we're doing our best" of quadrilaterals.

Memory Trick!

Parallelograms are the "steady Eddie" shapes — nothing fancy, but reliable.

6.2.2 Rectangle

All angles are right angles. This shape is the kid who follows all the rules and reminds the teacher to assign homework.

Memory Trick!

Rectangles = **right angles everywhere**. They're squares' calmer cousins.

6.2.3 Square

All sides equal, all angles right. The overachiever. The "golden child." The one grandparents brag about.

Memory Trick!

Squares are the "I did all the extra credit" shapes. Perfect sides, perfect angles — perfection everywhere.

6.2.4 Rhombus

All sides equal, but angles aren't necessarily right. It's like a square that stayed up too late and is a little crooked.

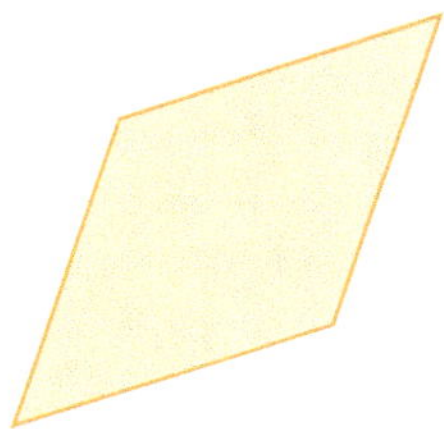

Memory Trick!

A rhombus is a square that partied too hard — same sides, wobblier angles.

6.2.5 Trapezoid

One pair of parallel sides. This shape is doing the bare minimum, and honestly, I respect that.

Memory Trick!

Trapezoids: one pair of parallel sides, zero guilt about it. They're the "I showed up, what more do you want?" shapes.

Chapter 7

Circles: The Shape That Never Ends (Like Your Questions)

Circles are beautiful. Perfect. Smooth. No corners. No drama. Honestly, if shapes were people, circles would be the chill friend who just wants everyone to get along.

7.1 Parts of a Circle

7.1.1 Center

The middle of the circle. Like the center of the living room where you kids leave all your socks.

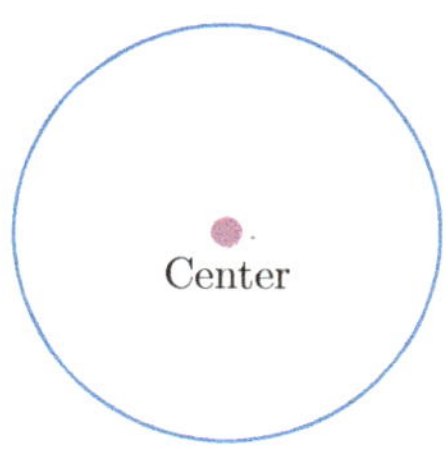

Memory Trick!

The center is the **heart of the circle**. Everything else is the same distance from it — very fair, very calm, very circle-like.

7.1.2 Radius

A segment from the center to the circle. It's the shortest path from "Dad, look!" to "Please stop climbing that."

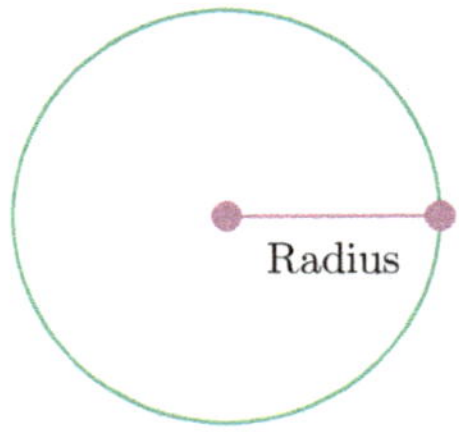

Memory Trick!

A radius is always **half the diameter**. If you remember that, circles suddenly get way easier.

7.1.3 Diameter

A segment passing through the center, touching both sides. Twice the radius. Twice the trouble.

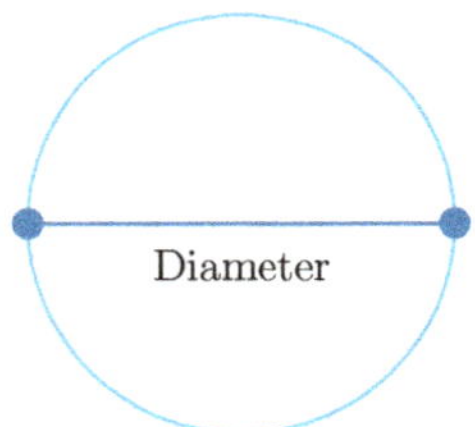

Memory Trick!

Diameter = **two radii back-to-back**. It's the longest segment you can draw inside a circle.

7.1.4 Chord

A segment connecting two points on the circle. Not as fancy as the diameter, but still important — like middle children.

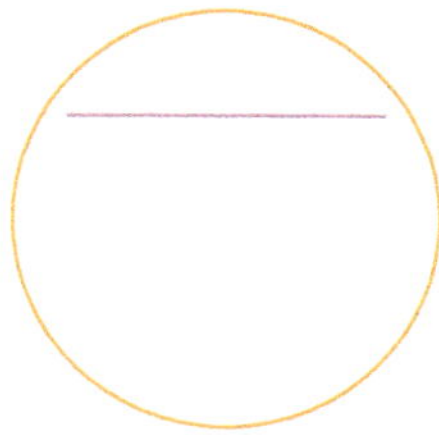

> **Memory Trick!**
>
> Every diameter is a chord, but not every chord is a diameter. It's like squares and rectangles all over again.

7.1.5 Arc

A piece of the circle's edge. Like the crust of a pizza slice — which you kids never eat.

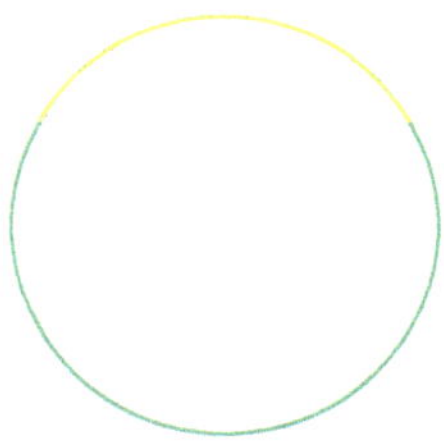

> **Memory Trick!**
>
> An arc is just a **curved slice** of the circle. If it looks like part of a pizza, you're on the right track.

7.2 Circumference

The distance around the circle. It's like walking around the house looking for where you left your phone.

$$C = 2\pi r$$

Memory Trick!

Circumference = **circle's perimeter**. If you know the radius, you know the whole story.

7.3 Area of a Circle

The space inside the circle. Imagine how much cereal you can fit in a circular bowl. (Too much. Always too much.)

$$A = \pi r^2$$

Memory Trick!

Area grows with the **square of the radius**. Double the radius? The area gets four times bigger. Wild, right?

Chapter 8

Polygons: Shapes With Many Sides and Even More Attitude

A polygon is any closed shape with straight sides. Some polygons are simple. Some are complicated. Some are like your homework — unnecessarily dramatic.

8.1 Regular vs. Irregular Polygons

8.1.1 Regular Polygon

All sides and angles are equal. This shape has its life together.

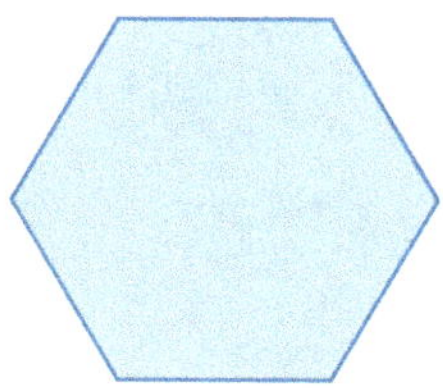

Memory Trick!

A regular polygon is the **"everything matches"** shape. Equal sides + equal angles = pure geometric harmony.

8.1.2 Irregular Polygon

Nothing matches. Nothing lines up. It's basically your bedroom.

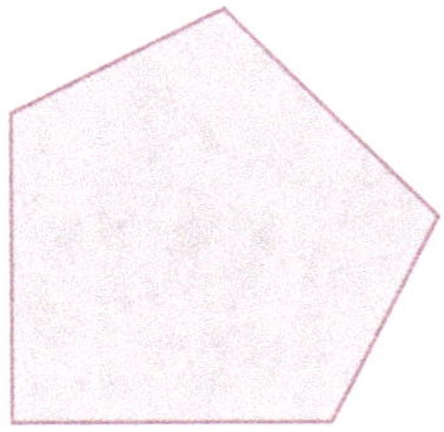

Memory Trick!

Irregular polygons are the **"I'll do what I want"** shapes. No equal sides required. No equal angles either.

8.2 Sum of Interior Angles

For any polygon with n sides:

$$\text{Sum of interior angles} = (n - 2) \times 180^\circ$$

Memory Trick!

Think of it like this: Every polygon can be split into $n - 2$ **triangles**. Each triangle has 180°, so multiply away.

Examples

- Triangle ($n = 3$): 180°
- Quadrilateral ($n = 4$): 360°
- Pentagon ($n = 5$): 540°
- Hexagon ($n = 6$): 720°

8.3 Exterior Angles

The sum of the exterior angles of ANY polygon is always:

$$360^\circ$$

Yes. Always. Even if the polygon looks like it was drawn by a toddler on a sugar rush.

Memory Trick!

Walk around any polygon, turning at each corner. By the time you get back to where you started, you've turned **one full circle** — 360°.

Chapter 9

Perimeter: Walking Around the Shape (Like I Do When I'm Looking for My Keys)

Perimeter is the distance around a shape. Think of it like this: if I walked around the entire house looking for where you kids hid the remote again, that walk would be the perimeter.

9.1 Perimeter of Common Shapes

9.1.1 Rectangle

$$P = 2(l + w)$$

Length plus width, doubled. Simple. Clean. Easier than getting you kids to agree on a movie.

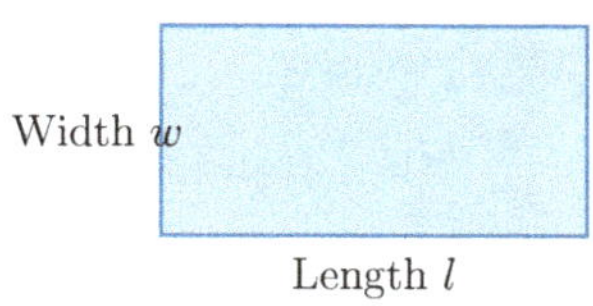

Memory Trick!

Rectangles have **two pairs of equal sides**. Add one length + one width, then double it — done.

9.1.2 Square

$$P = 4s$$

Four equal sides. Honestly, the square is the "I did all my homework early" of shapes.

Side s

Memory Trick!

Squares are the easiest perimeter ever: **four sides, all the same**. Just multiply by 4.

9.1.3 Triangle

Just add all three sides. No fancy formula. No drama. A rare moment of peace in geometry.

$$P = a + b + c$$

Memory Trick!

Triangles keep it simple: **three sides, one sum**. That's it.

9.2 Perimeter in Real Life

- Fencing a yard (so the dog stops escaping).
- Wrapping a present (badly, in my case).
- Measuring how far you walked while looking for your phone.

Memory Trick!

Perimeter = **the "walk around the outside" distance**. If you can walk it, you can measure it.

Chapter 10

Area: How Much Space a Shape Takes Up (Like Your Stuff All Over the Floor)

Area is the amount of space inside a shape. Imagine how much cereal fits in a bowl, or how much mess fits in your room. That's area.

Memory Trick!

Think of area as the **"fill it up"** measurement. If you can pour cereal into it, cover it with paint, or make a mess inside it, you're talking about area.

10.1 Area of Common Shapes

10.1.1 Rectangle

$$A = lw$$

Length times width. Even I can't mess that up.

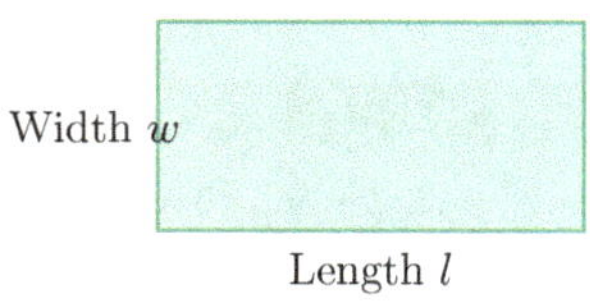

Memory Trick!

Rectangles are wonderfully predictable: **area = one side × the other side**. No tricks, no surprises.

10.1.2 Square

$$A = s^2$$

Side times itself. Like multiplying your chores by two when you argue.

Side s

Memory Trick!

Squares are rectangles with a perfection complex. **Same side × same side** — easy.

10.1.3 Triangle

$$A = \frac{1}{2}bh$$

Half the base times the height. Yes, half. Don't forget the half. I swear, if one of you forgets the half again...

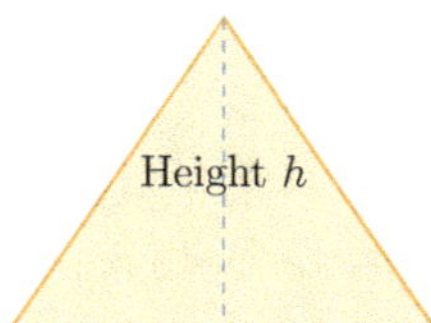

Base b

Memory Trick!

Triangles are basically rectangles cut in half. That's why the formula has a $\frac{1}{2}$. Forget the half, lose the points.

10.1.4 Parallelogram

$$A = bh$$

Same as a rectangle, but tilted. Like how you tilt your head when I try to explain taxes.

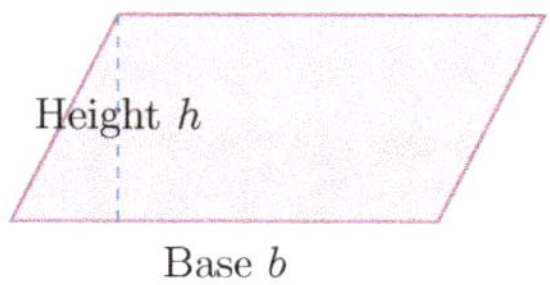

Memory Trick!

Parallelograms look slanted, but don't let them fool you. Their area is just **base** × **height**, same as rectangles.

10.1.5 Trapezoid

$$A = \frac{1}{2}(b_1 + b_2)h$$

Two bases, one height, and a formula that looks scarier than it is.

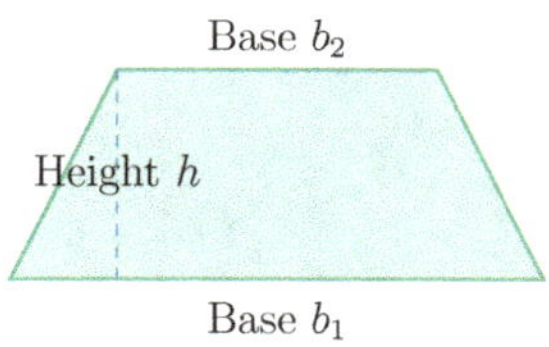

Memory Trick!

A trapezoid is like a triangle and a rectangle had a baby. Average the two bases, multiply by height — done.

10.1.6 Circle

$$A = \pi r^2$$

Pi times radius squared. And no, you can't replace π with 3. That's geometry blasphemy.

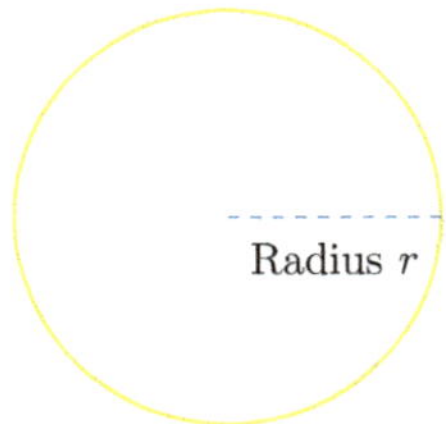

Memory Trick!

Circle area grows FAST. Double the radius? The area becomes **four times bigger**. Circles don't mess around.

Chapter 11

Surface Area: How Much Wrapping Paper You Need (And Why I Always Run Out)

Surface area is the total area of all the faces of a 3D shape. It's like wrapping a present — except geometry expects you to do it neatly.

Memory Trick!

Surface area = **every face added together**. If you'd have to cover it with wrapping paper, it counts.

11.1 Rectangular Prism

A rectangular prism is basically a fancy box. Like the ones you kids keep asking me to save "for crafts" and then never use.

$$SA = 2(lw + lh + wh)$$

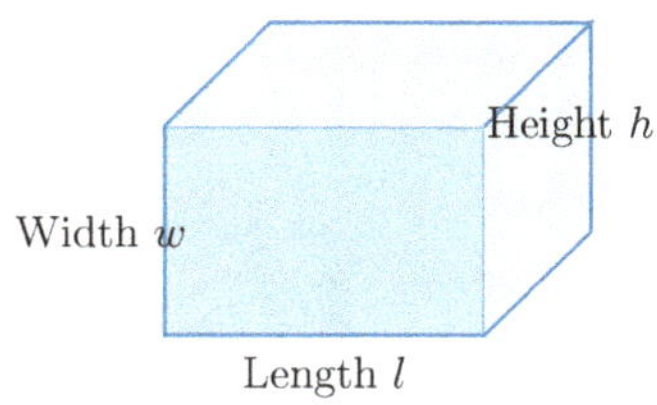

Memory Trick!

A rectangular prism has **three pairs of matching faces**. Find the area of each pair, double them, then add them up.

11.2 Cube

$$SA = 6s^2$$

Six identical square faces. Like six identical reminders to clean your room.

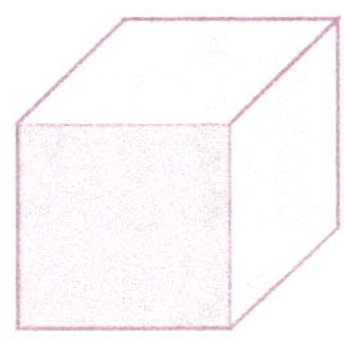

Memory Trick!

A cube is the easiest 3D shape ever: **one square × 6**. If you know the side length, you know everything.

11.3 Cylinder

$$SA = 2\pi r^2 + 2\pi rh$$

Two circles and a rectangle wrapped around. Like a can of soup, but with math.

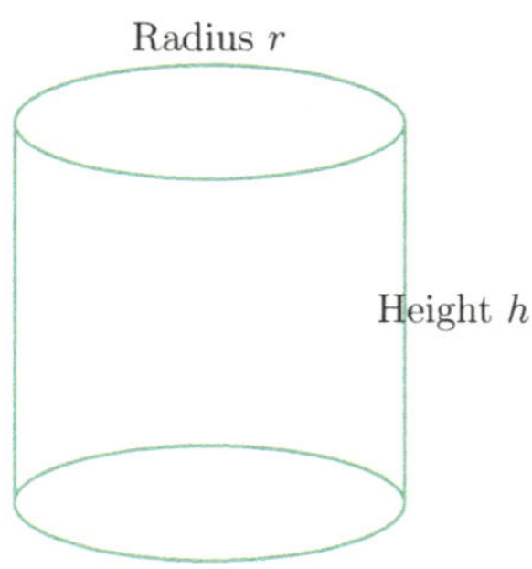

Memory Trick!

Cylinder surface area = **two circles + one big rectangle**. That rectangle is the "label" wrapped around the can.

Chapter 12

Volume: How Much Space Something Takes Up (Like Your Toys Everywhere)

Volume measures how much space a 3D object occupies. Think of it like this: if I filled your backpack with water — which I'm not saying I would, but I've thought about it — the amount of water it could hold is the volume.

Memory Trick!

Volume is the **"fill it up"** measurement. If you can pour something into it, stack something in it, or lose something inside it, you're talking about volume.

12.1 Rectangular Prism

$$V = lwh$$

Length times width times height. It's the "stack everything in the garage and hope it fits" formula.

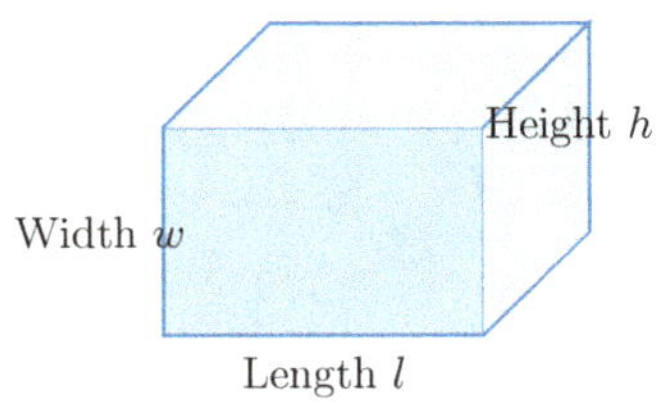

Memory Trick!

Rectangular prism volume is just **area of the base × height**. Build the base, stack it up — done.

12.2 Cube

$$V = s^3$$

Side times side times side. Like multiplying your chores by three when you "forget" them.

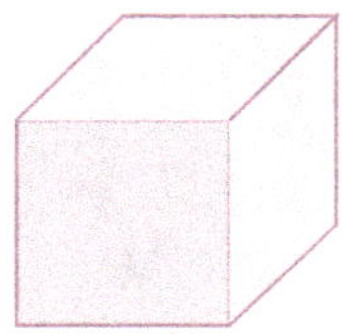

Memory Trick!

A cube is the "everything matches" box. **Side × side × side** — easy and satisfying.

12.3 Cylinder

$$V = \pi r^2 h$$

Circle area times height. Imagine stacking a bunch of circular pancakes. (Please don't actually stack pancakes on the table again.)

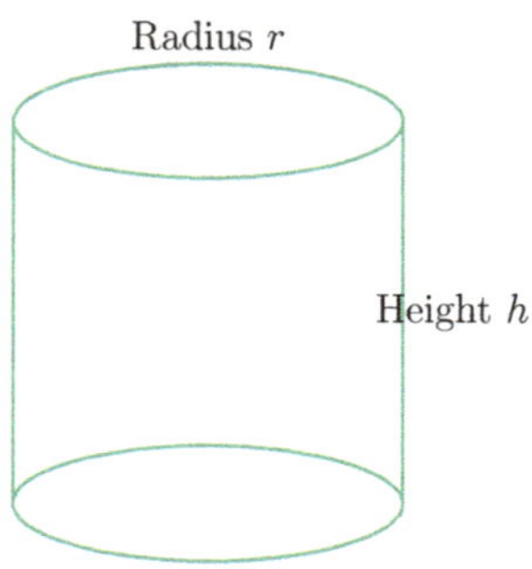

Memory Trick!

Cylinder volume = **area of the circle × height**. It's like stacking identical cookies into a tube.

12.4 Pyramid

$$V = \frac{1}{3}Bh$$

One-third the area of the base times the height. Why one-third? I don't know. Ask the ancient Egyptians.

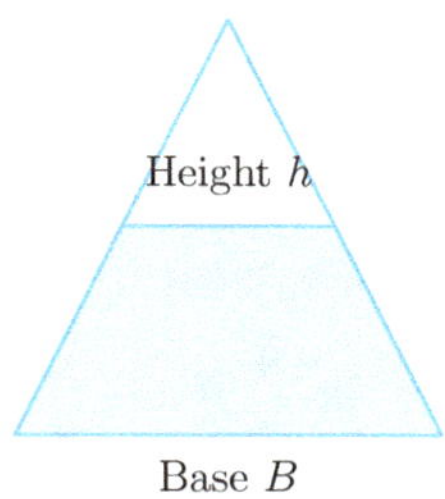

Memory Trick!

Pyramids take up **one-third** the space of a prism with the same base and height. It's geometry's way of keeping things interesting.

12.5 Cone

$$V = \frac{1}{3}\pi r^2 h$$

Same idea as the pyramid, but rounder. Like an ice cream cone — except this one won't melt all over the car.

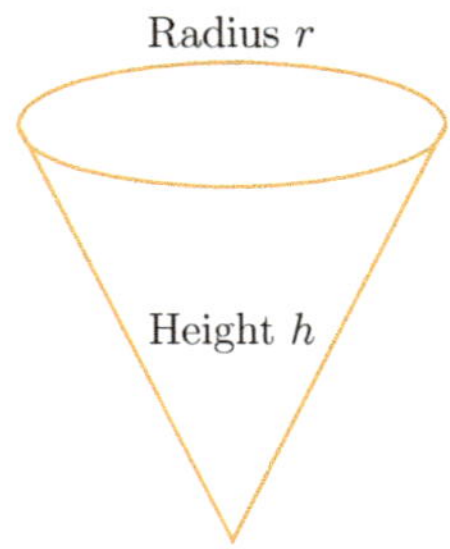

Memory Trick!

A cone is just a **circular pyramid**. Same formula, same one-third, just smoother.

Chapter 13

Coordinate Geometry: Plotting Points (And Plotting How to Get You Kids to Bed)

Coordinate geometry is where numbers and shapes meet. It's like when your worlds collide — math class meets art class.

13.1 The Coordinate Plane

The coordinate plane has two axes:

- The x-axis (horizontal)
- The y-axis (vertical)

They meet at the origin, which is the point $(0, 0)$. It's the "home base" of math — like the couch is home base for me.

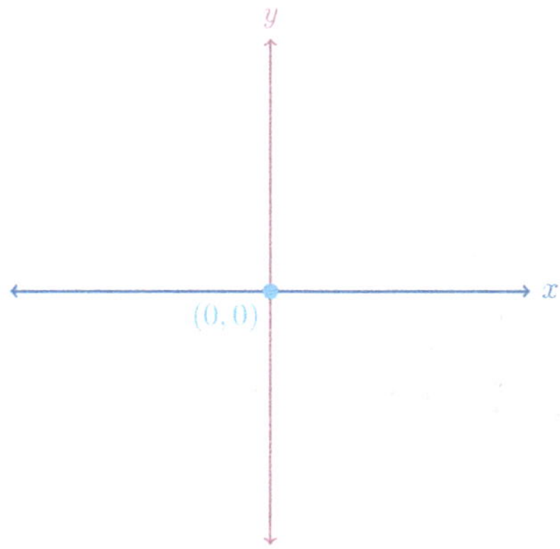

Memory Trick!

The origin is the **starting point of everything**. Every coordinate begins its journey from here.

13.2 Plotting Points

A point is written as (x, y).

- x tells you how far left or right.
- y tells you how far up or down.

It's like giving directions:

"Go 3 steps to the right, then 2 steps up. And please don't knock over the lamp this time."

> **Memory Trick!**
>
> Coordinates are just **directions on a grid**. Right/left first, then up/down — always in that order.

13.3 Distance Formula

To find the distance between two points:

$$d = \sqrt{(x_2 - x_1)^2 + (y_2 - y_1)^2}$$

Yes, it's the Pythagorean Theorem in disguise. Geometry loves disguises.

> **Memory Trick!**
>
> Distance = **hypotenuse of a right triangle**. You're secretly doing Pythagoras every time.

13.4 Midpoint Formula

To find the point exactly halfway between two points:

$$M = \left(\frac{x_1 + x_2}{2}, \frac{y_1 + y_2}{2} \right)$$

It's like splitting a cookie evenly. Which you kids never do.

Memory Trick!

Midpoint = **average the x's, average the y's**. That's it — the cookie-sharing formula.

Chapter 14

Transformations: Moving Shapes Around (Like You Move Your Stuff Everywhere)

Transformations change a shape's position, size, or orientation. Don't worry — the shape survives. Unlike my sanity when I step on a Lego at 2 a.m.

There are four main types of transformations, and each one messes with a shape in its own special way. Think of them like the four moods of geometry: sliding, flipping, spinning, and growing/shrinking.

—

14.1 Translation

A translation slides a shape from one place to another without turning it, flipping it, or stretching it. It's the laziest transformation — the shape just scoots over like you do when someone wants to sit next to you.

Every point in the shape moves the same distance in the same direction. No drama. No rotation. No resizing. Just a smooth slide.

Example: If a point moves from $(2, 3)$ to $(5, 7)$, then every point in the shape moves $+3$ in the x direction and $+4$ in the y direction.

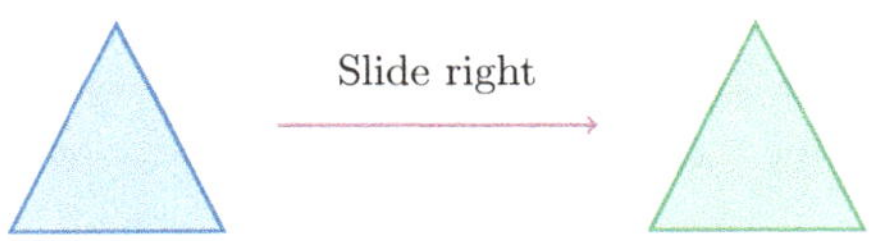

> **Memory Trick!**
>
> Translation = **same shape, same size, new place**. It's geometry's version of "scoot over."

—

14.2 Reflection

A reflection flips a shape over a line — the "mirror line." It's like the shape is checking itself out before picture day.

Left becomes right, right becomes left, and the whole shape reverses direction. But the size and angles stay exactly the same.

Example: Reflect the point $(4, 2)$ over the y-axis and it becomes $(-4, 2)$. Same height, opposite side.

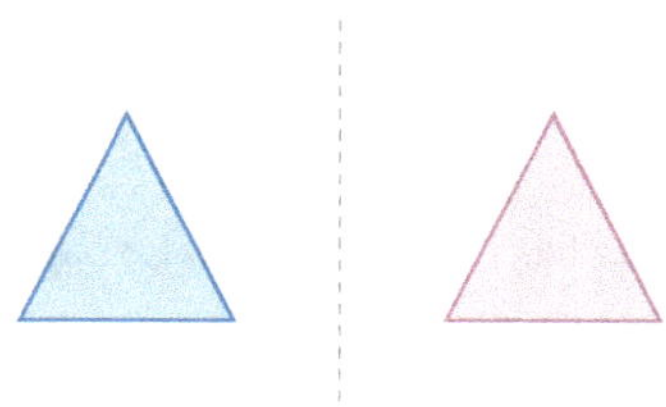

Mirror line

> **Memory Trick!**
>
> Reflection = **flip it**. Everything swaps sides, just like a mirror.

—

14.3 Rotation

A rotation turns a shape around a point — usually the origin, but it can be any point. The shape keeps its size and angles; it just spins around like you do when you're bored in the grocery store.

Rotations can be clockwise or counterclockwise, and they're measured in degrees.

Example: Rotating the point $(3, 0)$ by $90°$ counterclockwise around the origin sends it to $(0, 3)$.

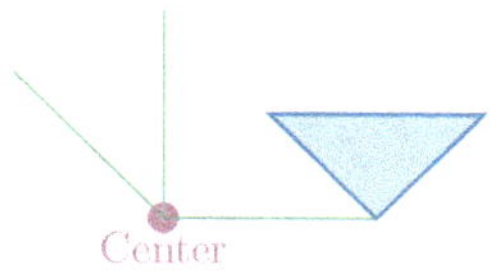

Memory Trick!

Rotation = **turn around a point**. Same shape, same size — just dizzy.

—

14.4 Dilation

A dilation changes the size of a shape but keeps its proportions the same. It's the "zoom in / zoom out" transformation — something you do on your phone way too often.

The scale factor tells you how much bigger or smaller the shape becomes: - Greater than $1 \to$ bigger - Between 0 and $1 \to$ smaller - Exactly $1 \to$ nothing changes (but nice try)

Example: A triangle with side lengths $3, 4, 5$ dilated by a scale factor of 2 becomes $6, 8, 10$.

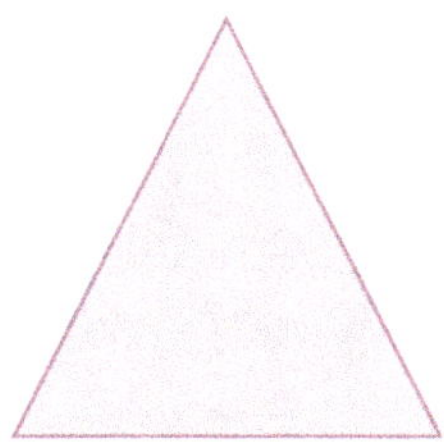

Scale factor 2

Memory Trick!

Dilation = **same shape, different size**. The scale factor is the boss.

Chapter 15

The Pythagorean Theorem: The One Formula You'll Never Forget (Because I Won't Let You)

Ah yes, the Pythagorean Theorem. The crown jewel of right triangles. The Beyoncé of geometry formulas.

If you have a right triangle — meaning one angle is exactly 90° — then the sides follow this magical relationship:

$$a^2 + b^2 = c^2$$

Where:

- a and b are the legs (the shorter sides)
- c is the hypotenuse (the long one opposite the right angle)

If you forget which side is the hypotenuse, look for the one that looks like it's trying the hardest. That's c — the overachiever.

Memory Trick!

The hypotenuse is always **across from the right angle**. If the triangle had a "main character," it'd be c.

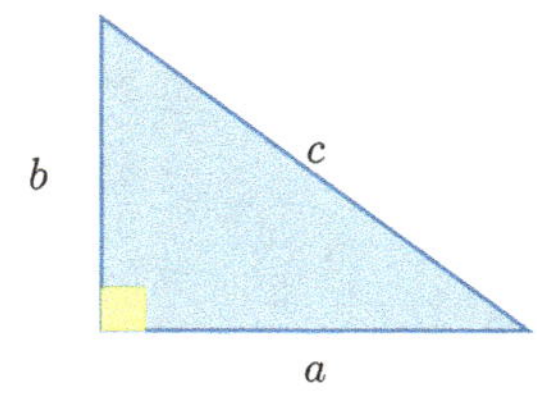

15.1 Examples

Example 1: The Classic 3-4-5 Triangle

$$3^2 + 4^2 = 9 + 16 = 25 = 5^2$$

This triangle is so famous it should have its own reality show.

Example 2: Dad's Ladder Problem

If Dad leans a 10-foot ladder against the house and the bottom is 6 feet away from the wall, how high does it reach?

$$6^2 + b^2 = 10^2$$

$$36 + b^2 = 100$$

$$b^2 = 64$$

$$b = 8$$

So the ladder reaches 8 feet high. And no, you may not climb it.

Memory Trick!

Whenever you see a right triangle in real life — ladders, ramps, shadows — the Pythagorean Theorem is quietly doing the math for you.

Chapter 16

Right Triangle Trigonometry: Sine, Cosine, and Dad Crying Quietly

Trigonometry sounds scary, but it's really just about ratios in right triangles. Ratios! That's it. If you can share a pizza fairly (you can't, but hypothetically), you can do trig.

16.1 SOH-CAH-TOA

This is the magic phrase that helps you remember:

$$\text{SOH: } \sin(\theta) = \frac{\text{opposite}}{\text{hypotenuse}}$$

$$\text{CAH: } \cos(\theta) = \frac{\text{adjacent}}{\text{hypotenuse}}$$

$$\text{TOA: } \tan(\theta) = \frac{\text{opposite}}{\text{adjacent}}$$

Memory Trick!

SOH-CAH-TOA is your **trig survival spell**. Say it three times fast and right triangles suddenly make sense.

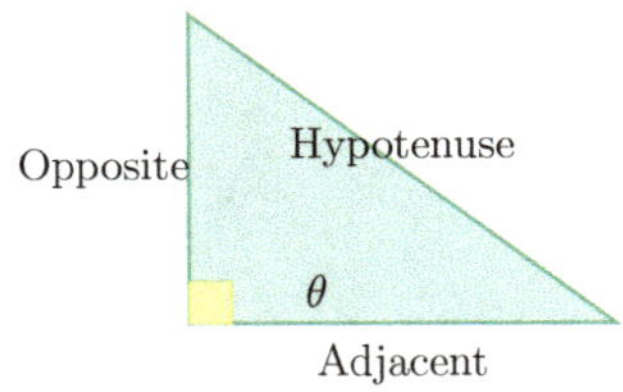

16.2 Why Trig Matters

- It helps you measure heights without climbing things.
- It's how engineers make sure bridges don't fall down.
- It's how I figured out the angle you kids throw your socks at.

Memory Trick!

Trig is everywhere — in architecture, physics, video games, and apparently in the trajectory of airborne socks.

Chapter 17

More Circle Theorems: Because Circles Refuse To Be Simple

Circles are like that one kid who always has follow-up questions. Just when you think you understand them, they say, "Actually, there's more."

17.1 Angles Formed Outside the Circle

When two lines intersect outside a circle, the angle formed is:

$$\text{Angle} = \frac{1}{2}(\text{Big arc} - \text{Small arc})$$

Yes, subtraction. Because circles want to keep you humble.

Memory Trick!

Outside-circle angles are **half the difference of the arcs**. Big arc minus small arc — it's like comparing chores: "What I *did*" minus "what you *said* I did."

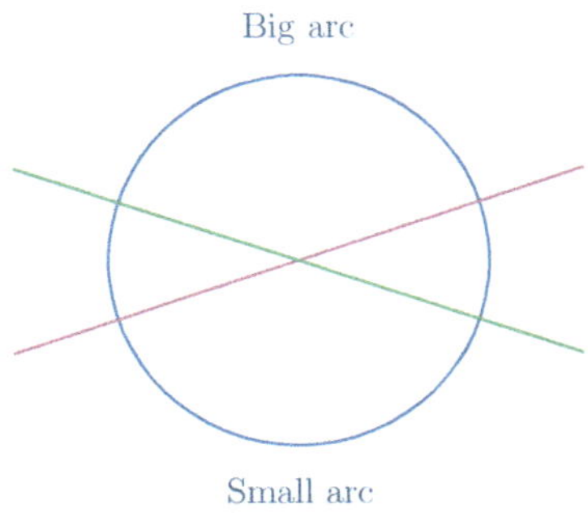

17.2 Tangents

A tangent touches a circle at exactly one point. Just one. Like how you kids "barely touched" the cookie jar.

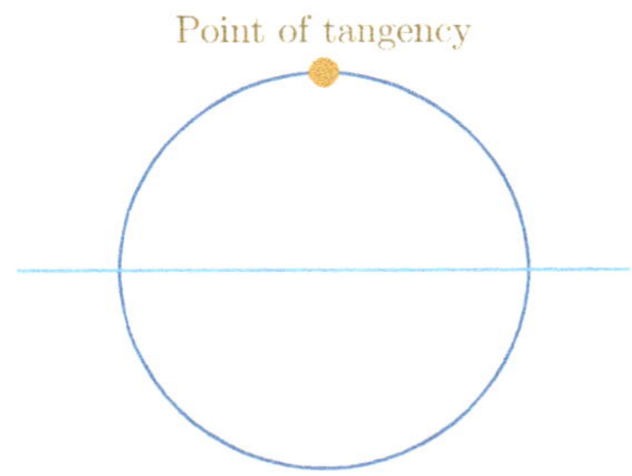

Important Tangent Facts

- A tangent is perpendicular to the radius at the point of tangency.
- Tangents from the same external point are equal in length.

Which means if you draw two tangents from the same point, they're twins. Not identical-twins-who-switch-places-to-confuse-me twins, but close.

Memory Trick!

Tangents are **perfectly behaved**: they meet the circle once, form right angles with the radius, and come in matching pairs from the same point.

Chapter 18

3D Geometry: Shapes That Pop Out of the Page (Like You Kids at Bedtime)

Welcome to the world of 3D shapes — prisms, pyramids, cylinders, cones, and spheres. These shapes have volume, surface area, and a special talent for making homework take longer than expected.

18.1 Nets of 3D Shapes

A net is a 2D pattern you can fold into a 3D shape. It's like origami, but with more math and fewer paper cranes.

18.1.1 Cube Net

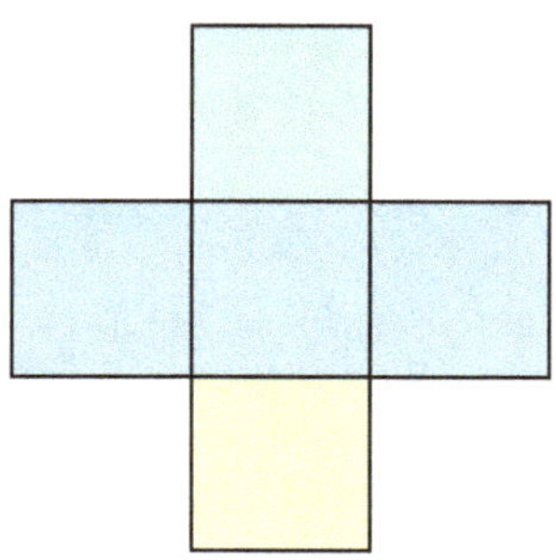

If you fold this up, you get a cube. If you fold it wrong, you get... something else. (And that's okay. Dad believes in you.)

Memory Trick!

A net is the **unfolded version** of a 3D shape. If it looks like a weird Tetris piece, you're probably doing it right.

18.2 Cross Sections

A cross section is what you get when you slice a 3D shape. Like cutting a cake — except this time you're allowed to.

- Slice a cylinder horizontally → circle
- Slice it vertically → rectangle
- Slice it diagonally → ellipse (fancy!)

Memory Trick!

Cross sections show you the **hidden shapes inside** 3D objects. It's geometry's version of "what's really going on in there?"

18.3 Spheres

A sphere is a perfectly round 3D shape. Like a basketball. Or the meatballs you kids keep stealing from the pot.

$$V = \frac{4}{3}\pi r^3$$

$$SA = 4\pi r^2$$

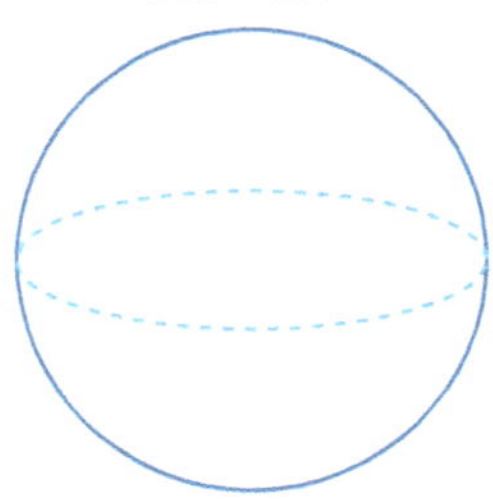

Sphere

Memory Trick!

Spheres are the **smoothest shapes in geometry**. No edges, no corners — just pure roundness.

Chapter 19

Similarity in 3D: Big Shapes, Little Shapes, Same Vibes

Just like 2D shapes, 3D shapes can be similar too. They're like siblings — same shape, different size, same ability to argue loudly. The only difference is that 3D shapes take up actual space, which means their scaling gets a little more dramatic.

When two 3D shapes are similar, every linear measurement grows (or shrinks) by the same scale factor. But areas and volumes? Oh, they grow faster than your laundry pile on a Friday.

19.1 Scale Factor

If the scale factor is k:

- Linear dimensions multiply by k
- Surface area multiplies by k^2
- Volume multiplies by k^3

Yes, cubes everywhere. Geometry loves cubes almost as much as you kids love snacks.

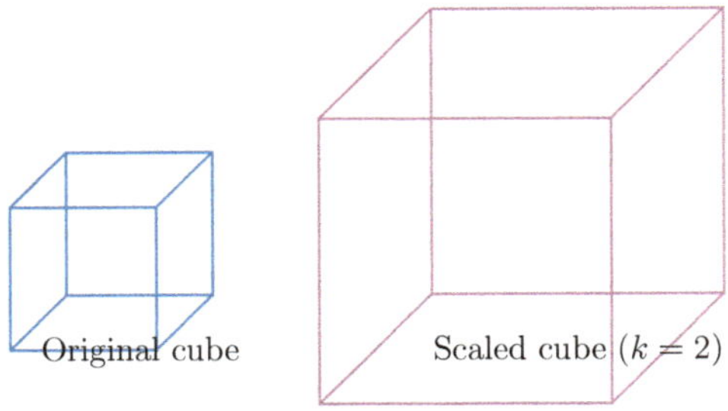

Memory Trick!

When shapes scale up or down, **lengths grow by k, areas by k^2, and volumes by k^3**. It's geometry's version of "things escalate quickly."

Example

If a cube's side length doubles:

- Side length $\rightarrow 2s$
- Surface area $\rightarrow$ 4 times bigger
- Volume $\rightarrow$ 8 times bigger

It grows faster than your laundry pile. And that's saying something.

—

Chapter 20

Coordinate Geometry With Shapes: Dad's Favorite Part (Don't Tell the Others)

Let's put shapes on the coordinate plane and see what happens. This is where algebra and geometry shake hands and pretend they get along.

20.1 Finding the Area of a Triangle Using Coordinates

There's a fancy formula for this:

$$A = \frac{1}{2}\left|x_1(y_2 - y_3) + x_2(y_3 - y_1) + x_3(y_1 - y_2)\right|$$

Yes, it looks scary. Yes, it works. Yes, I'll help you through it.

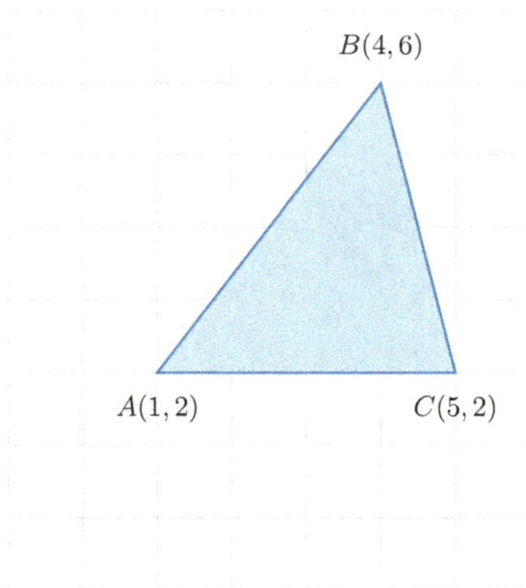

Memory Trick!

This formula is basically **area = base × height ÷ 2**, but dressed up in coordinate-plane clothing.

Example

Plug into the formula:

$$A = \frac{1}{2}|1(6-2) + 4(2-2) + 5(2-6)|$$

$$A = \frac{1}{2}|4 + 0 - 20|$$

$$A = \frac{1}{2}|-16| = 8$$

Boom. Triangle area = 8 square units. Dad victory dance. (You're lucky I'm not demonstrating it.)

—

Chapter 21

Vectors: Arrows With Attitude

A vector has direction and magnitude. It's like when you kids run across the room — fast, chaotic, and in a very specific direction.

21.1 Vector Basics

A vector is written like this:

$$\vec{v} = \langle a, b \rangle$$

Where:

- a is the horizontal change
- b is the vertical change

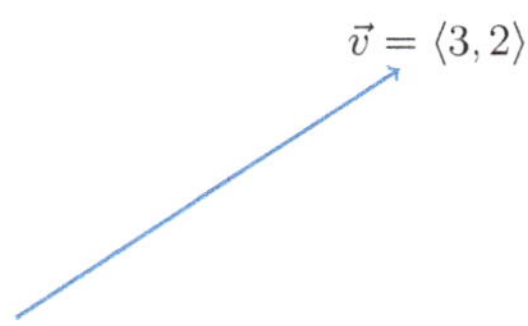

Memory Trick!

A vector is just a **movement**: "Go a to the side, then b up or down." It's math's version of giving directions.

21.2 Adding Vectors

$$\langle a, b \rangle + \langle c, d \rangle = \langle a + c,\ b + d \rangle$$

Just add the components. Like adding chores to your list — one more thing on top of another.

Example:

$$\langle 3,2\rangle + \langle -1,4\rangle = \langle 2,6\rangle$$

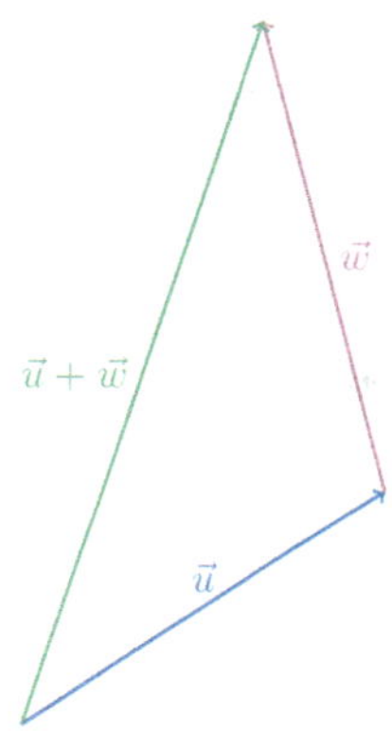

Memory Trick!

Vector addition = **combine the movements**. Where you end up is the sum of where you were going.

21.3 Vector Diagrams

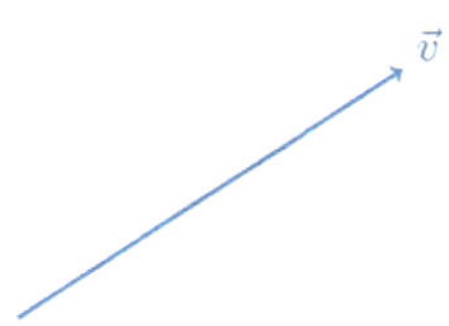

Memory Trick!

A vector diagram is just an **arrow showing direction and length**. Long arrow = big magnitude. Short arrow = "barely moved."

21.4 Magnitude of a Vector

$$|\vec{v}| = \sqrt{a^2 + b^2}$$

Yes, it's the Pythagorean Theorem again. It's everywhere. Like socks on the floor.

Memory Trick!

Magnitude = **how long the arrow is**. And yes, Pythagoras shows up here too — he's unavoidable.

Chapter 22

More Vector Fun: Because Arrows Deserve Attention Too

Vectors aren't just arrows — they're tiny stories. They tell you where something is going and how fast it's getting there. Kind of like you kids sprinting toward the ice cream truck.

22.1 Subtracting Vectors

Subtracting vectors is just like adding them... except one of them is grumpy. You flip its signs, and suddenly the whole direction changes.

$$\langle a, b\rangle - \langle c, d\rangle = \langle a - c,\ b - d\rangle$$

Example:

$$\langle 4, 3\rangle - \langle 1, -2\rangle = \langle 4 - 1,\ 3 - (-2)\rangle = \langle 3,\ 5\rangle$$

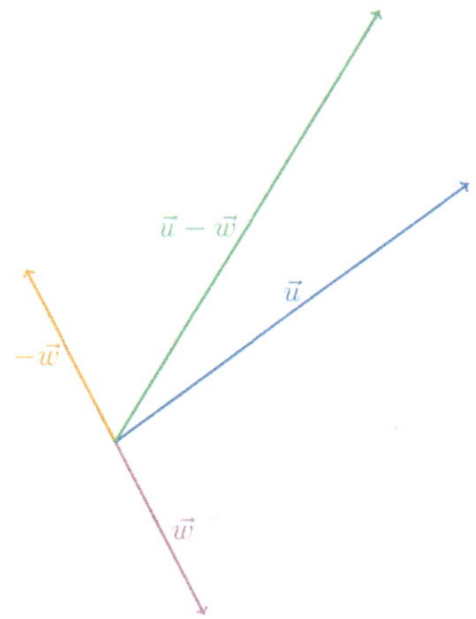

It's like taking away screen time. One minute you have it, the next minute... gone.

> **Memory Trick!**
>
> Subtracting vectors is just **adding the opposite**. Flip the signs of the second vector and add — easy.

—

22.2 Scalar Multiplication

Multiply a vector by a number and it gets longer or shorter. Same direction, different length — like your energy level before and after sugar.

$$k \cdot \langle a, b \rangle = \langle ka,\ kb \rangle$$

Example:

$$3 \cdot \langle 2, 1 \rangle = \langle 6, 3 \rangle$$

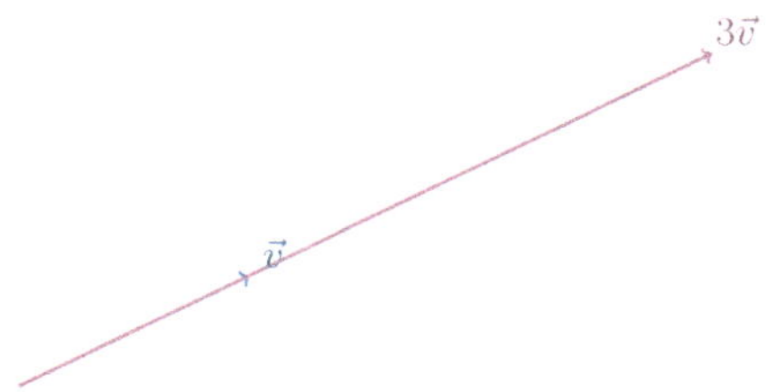

Memory Trick!

Scalar multiplication = **stretching or shrinking**. Same direction, new length — like zooming in or out.

—

22.3 Unit Vectors

A unit vector has length 1. It's the "I'm just a little guy" of vectors.

$$\hat{v} = \frac{\vec{v}}{|\vec{v}|}$$

Example:

$$\vec{v} = \langle 6, 8 \rangle, \quad |\vec{v}| = 10$$

$$\hat{v} = \left\langle \frac{6}{10}, \frac{8}{10} \right\rangle = \langle 0.6,\ 0.8 \rangle$$

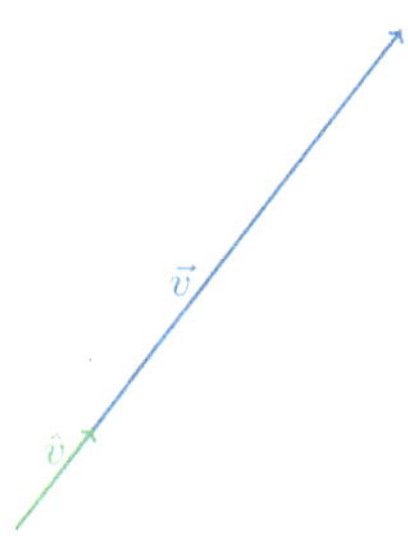

Useful for direction. Not useful for deciding what's for dinner.

Memory Trick!

Unit vectors are the **pure direction** of a vector. They point the way without adding extra length.

Chapter 23

Logic in Geometry: Because Shapes Don't Lie (But You Kids Sometimes Do)

Geometry uses logic — real logic — not the kind you kids use when you say, "Technically, I didn't touch the cookie. I just held it."

23.1 Conditional Statements

A conditional statement has the form:

If p, then q.

Example:
"If it's raining, then recess is inside."

If it rains $\longrightarrow$ Recess is inside

Memory Trick!

Conditional statements are **cause** $\rightarrow$ **effect**. If the first thing happens, the second thing must follow.

—

23.2 Converse, Inverse, Contrapositive

- **Converse:** If q, then p.

- **Inverse:** If not p, then not q.
- **Contrapositive:** If not q, then not p.

Only the contrapositive is always true when the original is true. The others are like your excuses — sometimes true, sometimes not.

Original

Converse

Memory Trick!

Contrapositive = **the reliable sibling**. If the original is true, the contrapositive is always true too.

—

23.3 Biconditional

$$p \leftrightarrow q$$

Means "if and only if." Like "You can have dessert if and only if you finish your homework."

Memory Trick!

Biconditional means **both directions work**. Each statement guarantees the other.

—

Chapter 24

Proofs With Parallel Lines: More Drama, More Angles

Let's prove some things using parallel lines. Because nothing says "family bonding" like angle relationships.

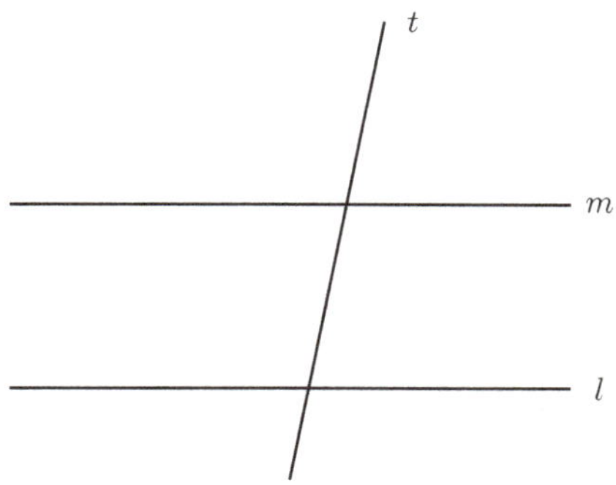

Parallel lines create angle relationships faster than siblings create arguments.

Memory Trick!

Parallel lines create **predictable angle pairs**. Once you spot the transversal, the angle relationships fall into place.

24.1 Example Proof

Before we jump into the formal proof, let's actually *see* what's going on. Parallel lines behave like well-trained siblings — predictable, consistent, and occasionally dramatic.

When a transversal cuts across two parallel lines, it creates angle pairs that match up perfectly. Corresponding angles are one of those pairs: same relative position, same measure, same "vibe."

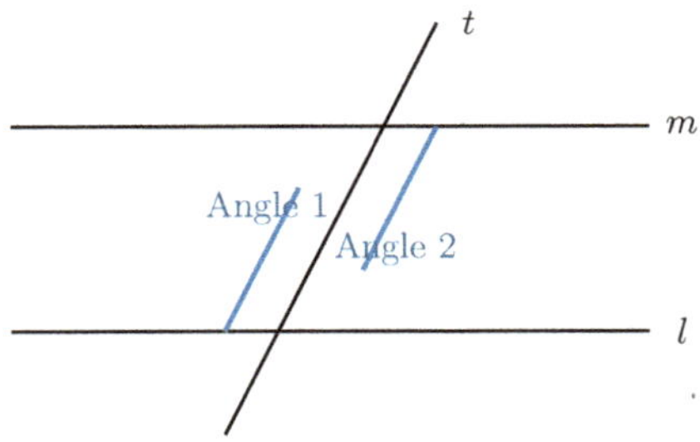

Angle 1 and Angle 2 are corresponding angles — same corner, different floor.

Now the formal proof:

Given:

Lines l and m are parallel.

Prove:

Corresponding angles are congruent.

Statements	Reasons
1. $l \parallel m$	Given
2. Transversal t intersects l and m	Definition of transversal
3. Corresponding angles form	Definition of corresponding angles
4. Corresponding angles are congruent	Parallel lines theorem

Boom. Done. Dad needs a snack.

Memory Trick!

Parallel lines create **predictable angle pairs**. Once you spot the transversal, the angle relationships fall into place.

Chapter 25

Area of Regular Polygons: Fancy Shapes, Fancy Formulas

Regular polygons are the overachievers of geometry. All sides equal. All angles equal. They probably color-code their notes and alphabetize their snacks.

25.1 Area Formula

To find the area of a regular polygon, we use:

$$A = \frac{1}{2}aP$$

Where:

- a = apothem (distance from center to a side)
- P = perimeter

The apothem is like the polygon's "height" for each triangular slice. Yes, "apothem" is a real word. No, it's not a type of medicine (though it sounds like one).

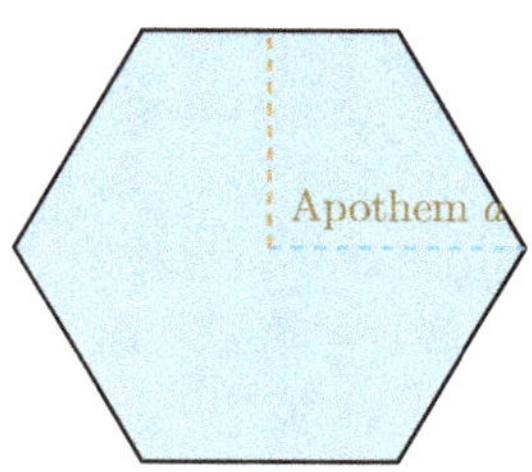

Memory Trick!

Area of a regular polygon = $\frac{1}{2}$ × **apothem** × **perimeter**. It's like slicing the shape into identical pizza slices and adding them up.

25.2 Example

Hexagon with side length 4:

$$P = 6 \times 4 = 24$$

Apothem (approx):

$$a \approx 3.46$$

$$A = \frac{1}{2}(3.46)(24) \approx 41.5$$

Nice. Clean. Satisfying. Like when you finally clean your room and I can see the floor again.

Chapter 26

Special Right Triangles: The Ones You Should Memorize

These triangles show up everywhere — tests, homework, real life, and probably your dreams if you study too late.

26.1 The 45-45-90 Triangle

$$\text{Leg} = x, \quad \text{Leg} = x, \quad \text{Hypotenuse} = x\sqrt{2}$$

It's the "twin triangle." Both legs are equal — like two kids who insist they're the same height.

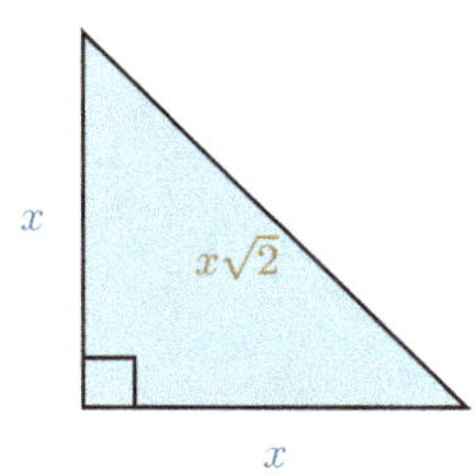

Memory Trick!

The 45-45-90 triangle is **perfectly balanced**. Two equal legs, one hypotenuse that's $x\sqrt{2}$ — easy to memorize.

26.2 The 30-60-90 Triangle

$$\text{Short leg} = x, \quad \text{Long leg} = x\sqrt{3}, \quad \text{Hypotenuse} = 2x$$

This triangle is dramatic. One short side, one long side, one hypotenuse that thinks it's better than everyone else.

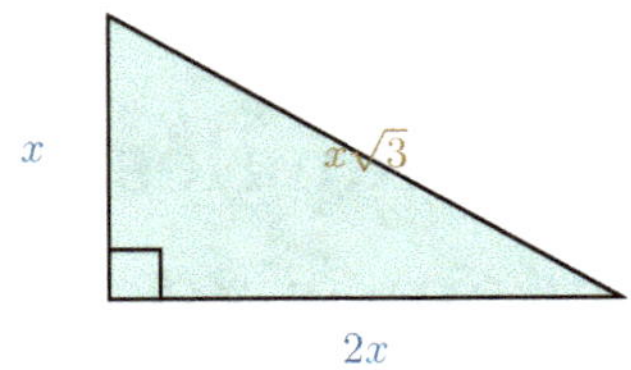

Memory Trick!

The 30-60-90 triangle is **the overachiever**. One tiny leg, one tall leg, and a hypotenuse that doubles the short side.

Chapter 27

Transformations on the Coordinate Plane: Moving Shapes With Math Magic

Let's combine transformations with coordinates. Because why not make things more interesting?

27.1 Translations

$$(x, y) \to (x + a,\ y + b)$$

Move right a, up b. Like dragging your backpack instead of lifting it.

Memory Trick!

Translation = **slide**. Nothing rotates, nothing flips — it just moves.

27.2 Reflections

$$(x, y) \to (-x, y) \quad \text{over the } y\text{-axis}$$

$$(x, y) \to (x, -y) \quad \text{over the } x\text{-axis}$$

Flip the shape like a pancake. (But please don't flip actual pancakes onto the ceiling again.)

Memory Trick!

Reflection = **mirror image**. Left becomes right, up becomes down.

27.3 Rotations

$$(x, y) \to (-y, x) \quad 90^\circ$$

$$(x, y) \to (-x, -y) \quad 180^\circ$$

$$(x, y) \to (y, -x) \quad 270^\circ$$

Spin the shape around the origin. Like you spinning in circles when you're bored.

Memory Trick!

Rotation = **turn around the origin**. Each rule is just a clever coordinate swap.

27.4 Dilations

Dilations are the "zoom in / zoom out" button of geometry. They change the size of a shape but keep the same proportions — which is more than I can say for the proportions of the laundry pile in this house.

$$(x, y) \to (kx,\ ky)$$

Every point gets multiplied by the same scale factor k. No bending, no twisting, no flipping — just a clean stretch or shrink.

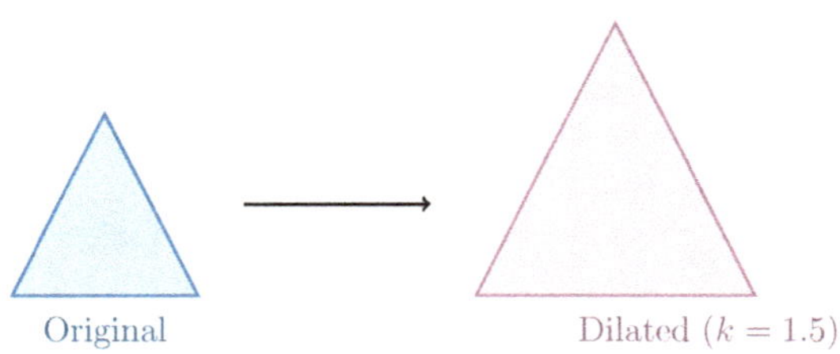

Scale the shape. Make it bigger. Make it smaller. Geometry is flexible — unlike my back.

Memory Trick!

Dilation = **multiply every coordinate by the scale factor**. Same shape, new size — simple and satisfying.

Chapter 28

More on Dilations: Making Shapes Grow (Like You Kids) or Shrink (Like My Patience)

Dilations change the size of a shape but keep the same proportions. It's like zooming in or out — except you can't pinch the paper.

28.1 Scale Factor

The scale factor k tells you how much bigger or smaller the shape gets.

- $k > 1 \rightarrow$ shape gets bigger
- $0 < k < 1 \rightarrow$ shape gets smaller
- $k = 1 \rightarrow$ shape stays the same

If only I could set $k = 0$ on the noise level in this house.

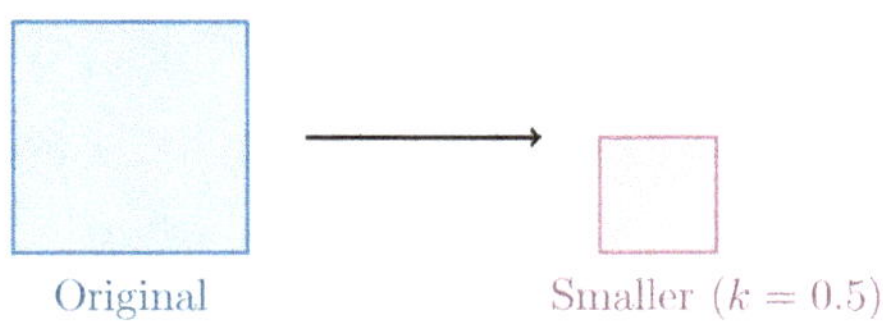

Memory Trick!

Dilations = **multiply every coordinate by k.** That's it. Same shape, new size — like magic, but with math instead of glitter.

28.2 Example

Triangle with vertices:

$$A(1,2),\ B(3,4),\ C(5,1)$$

Dilate by $k = 2$:

$$A'(2,4),\ B'(6,8),\ C'(10,2)$$

Everything doubles. Except your willingness to do chores.

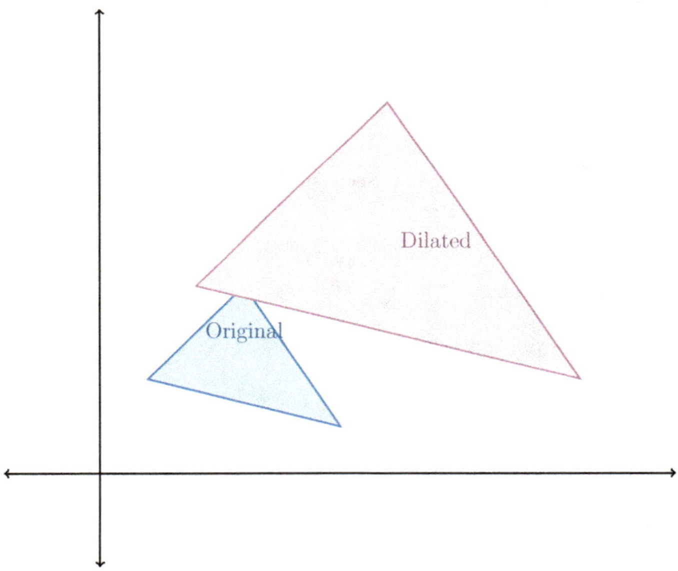

Chapter 29

Distance, Midpoint, and Slope: The Coordinate Plane Triple Threat

These three formulas show up everywhere — tests, homework, and that one time I tried to build a treehouse and realized I should've measured things first.

29.1 Distance Formula

$$d = \sqrt{(x_2 - x_1)^2 + (y_2 - y_1)^2}$$

Yes, it's the Pythagorean Theorem again. It's basically the guest who never leaves.

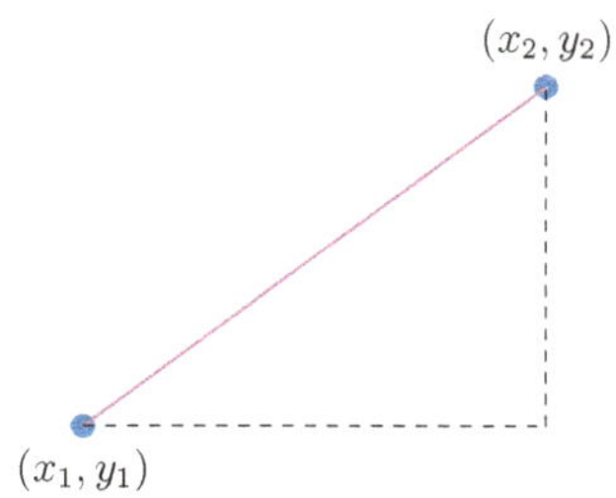

Memory Trick!

Distance = **length of the segment**. You're secretly doing Pythagoras every time you measure it.

29.2 Midpoint Formula

$$M = \left(\frac{x_1 + x_2}{2}, \frac{y_1 + y_2}{2}\right)$$

Perfect for splitting things evenly. Which you kids never do.

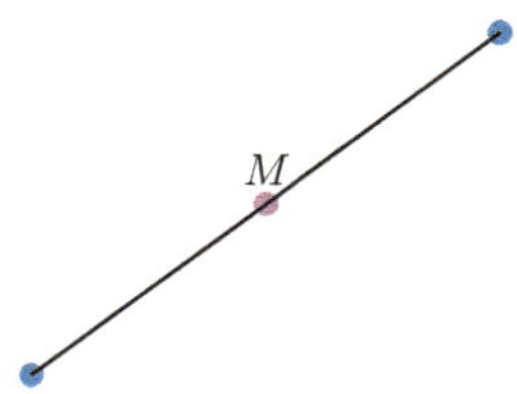

Memory Trick!

Midpoint = **average the x's, average the y's**. It's the "fair share" formula — in theory.

29.3 Slope Formula

$$m = \frac{y_2 - y_1}{x_2 - x_1}$$

Slope tells you how steep a line is. Like the hill I have to walk up when you forget your backpack at school.

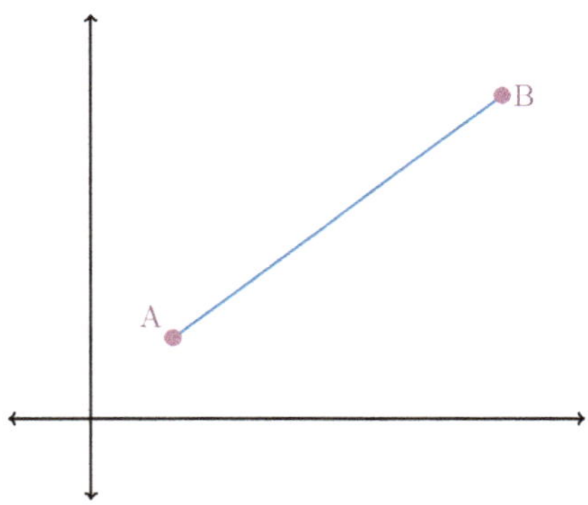

Memory Trick!

Slope = **rise over run**. How much up vs. how much over — simple and powerful.

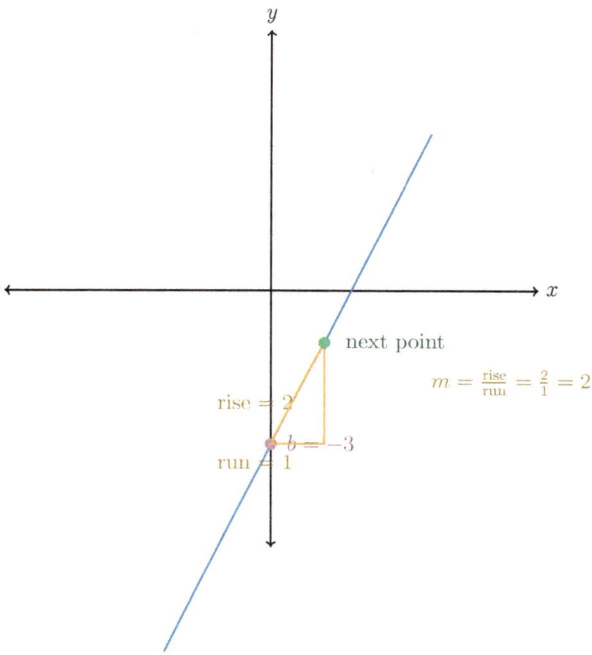

29.4 Example 7: Circles

A circular garden has radius 6 feet. Find the area.

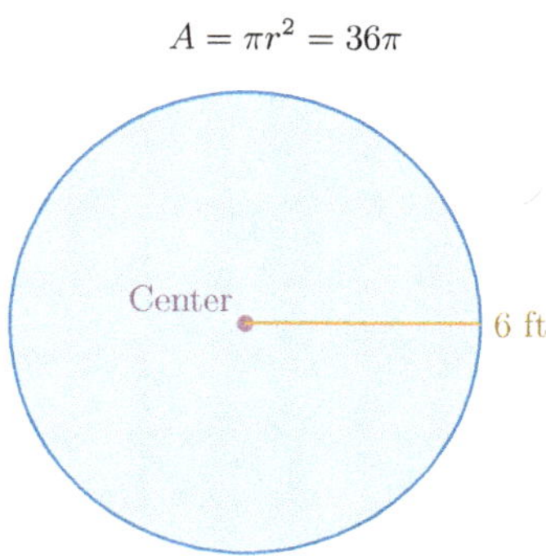

Chapter 30

Geometry in Real Life: Yes, You Actually Use This Stuff

You might think geometry is just shapes on paper, but it's everywhere. It sneaks into your house, your hobbies, your snacks, your sports — geometry is basically the quiet roommate who pays rent by showing up in everything you do.

30.1 Architecture

Buildings use:

- Triangles for support
- Circles for arches
- Rectangles for rooms

Triangles are the unsung heroes of architecture — they don't wobble, they don't complain, they just hold everything up. Circles make arches strong and pretty. Rectangles make rooms easy to build... and even easier to clutter.

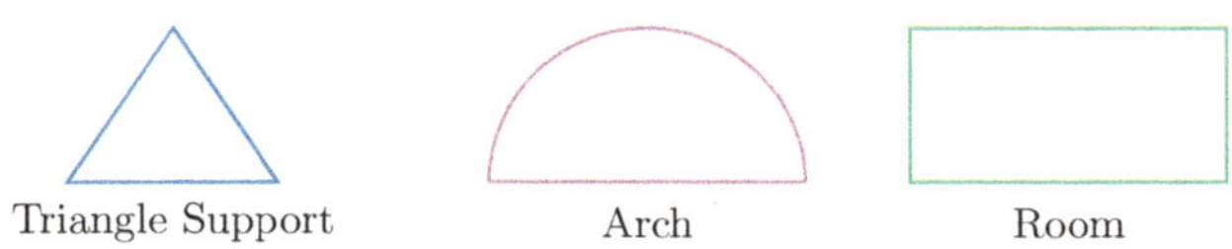

And sometimes weird shapes that make Dad say, "Who designed this hallway?"

Memory Trick!

Architecture is geometry wearing a hard hat. Every beam, arch, and window has math behind it.

30.2 Sports

- Basketball arcs
- Soccer angles
- Baseball trajectories

Every sport is basically a geometry problem pretending to be fun.

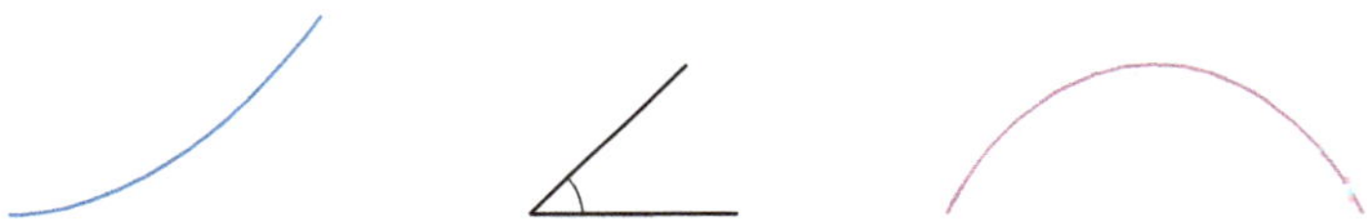

Basketball Arc   Soccer Angle   Baseball Path

Geometry is basically the secret coach of every sport.

30.3 Art

Artists use:

- Symmetry
- Proportion
- Perspective

Even your doodles have geometry — yes, even the ones on your homework.

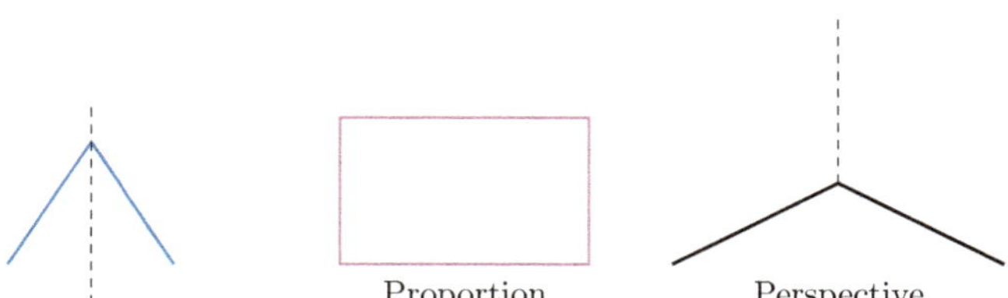

Symmetry   Proportion   Perspective

30.4 Nature

- Honeycombs → hexagons
- Snowflakes → symmetry
- Tree branches → fractals

Nature is basically a giant geometry textbook.

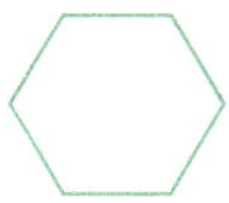

Hexagon

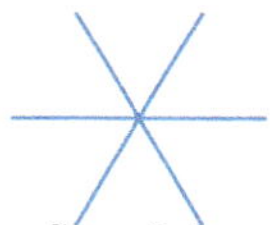

Snowflake

Fractal Branching

Chapter 31

Fractals: Shapes That Get Weird (But in a Cool Way)

Fractals are shapes that repeat over and over at smaller and smaller scales. Like when you kids repeat the same joke until I pretend to laugh.

31.1 The Koch Snowflake

Start with a triangle. Then add smaller triangles. Then add even smaller triangles. Repeat until your brain melts.

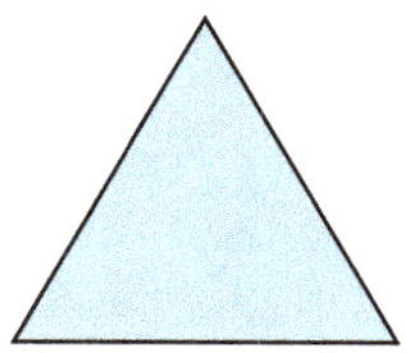

(We're not drawing the full fractal here — Dad's tired.)

31.2 The Sierpinski Triangle

Take a triangle. Remove the middle. Repeat forever.

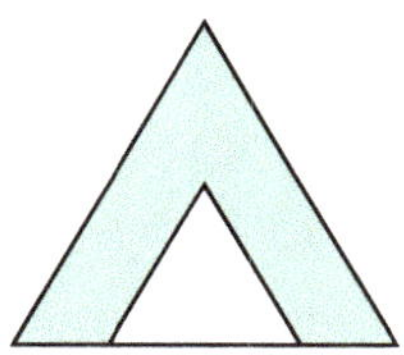

It's like geometry meets Minecraft.

Memory Trick!

Fractals = **patterns inside patterns**. Zoom in forever and the shape keeps repeating.

Chapter 32

Probability and Geometry: When Shapes Get Random

Sometimes geometry and probability team up. It's like when two of you kids team up — things get interesting fast.

32.1 Geometric Probability

$$\text{Probability} = \frac{\text{Favorable area}}{\text{Total area}}$$

Example

A dartboard has radius 10. The bullseye has radius 2. What's the probability of hitting the bullseye?

$$\text{Bullseye area} = \pi(2)^2 = 4\pi$$

$$\text{Total area} = \pi(10)^2 = 100\pi$$

$$P = \frac{4\pi}{100\pi} = \frac{1}{25}$$

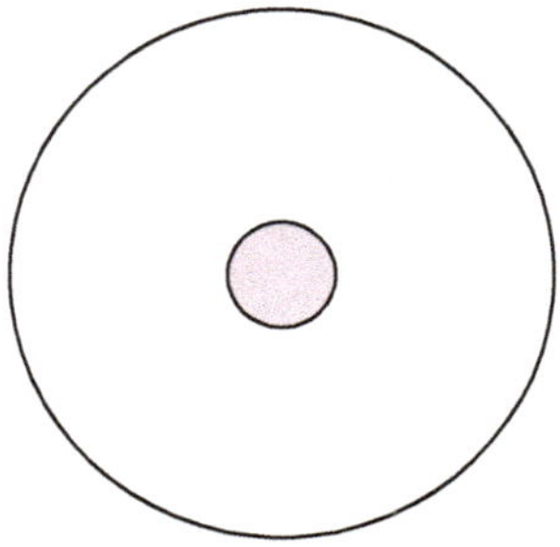

Dartboard

So... not great odds. Kind of like the odds of you kids remembering to put your dishes away.

Memory Trick!

Geometric probability = **area of target ÷ total area**. Bigger target = better odds. Life lesson right there.

Chapter 33

Transformations Review: Because Dad Knows You Forgot

Let's recap the big four transformations:

33.1 Translation

Slide the shape. No turning. No flipping. Just sliding — like socks on hardwood floors.

33.2 Reflection

Flip the shape over a line. Like flipping a pancake, but less messy.

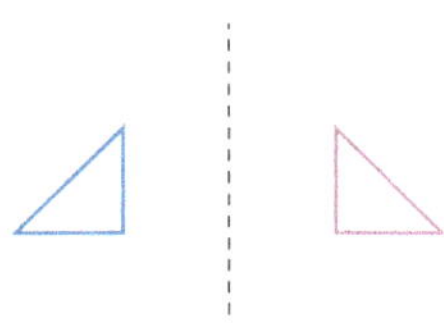

33.3 Rotation

Turn the shape around a point. Like you spinning in circles when you're bored.

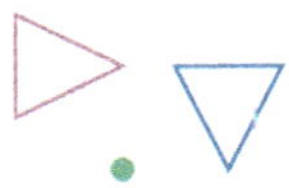

33.4 Dilation

Resize the shape. Bigger or smaller — like your appetite depending on what's for dinner.

Memory Trick!

Transformations = **move it, flip it, spin it, shrink/grow it**. Everything else is just details.

Chapter 34

Final Review: Everything You Need To Know (Before Dad Falls Asleep)

Let's wrap this up with a big review. If you remember these, you'll be ready for anything geometry throws at you.

34.1 The Big Ideas

Points, Lines, Segments, Rays

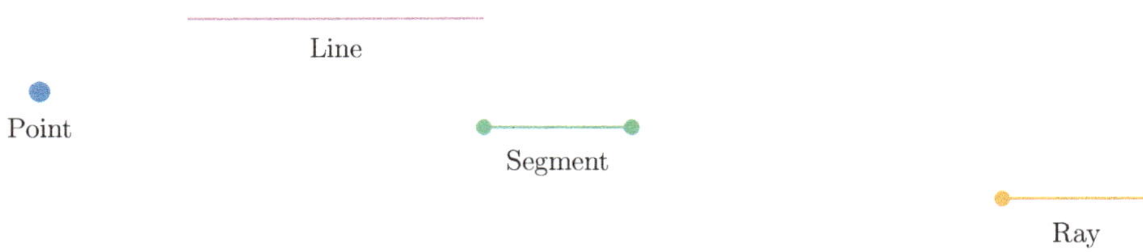

Angles and Their Types

Triangles

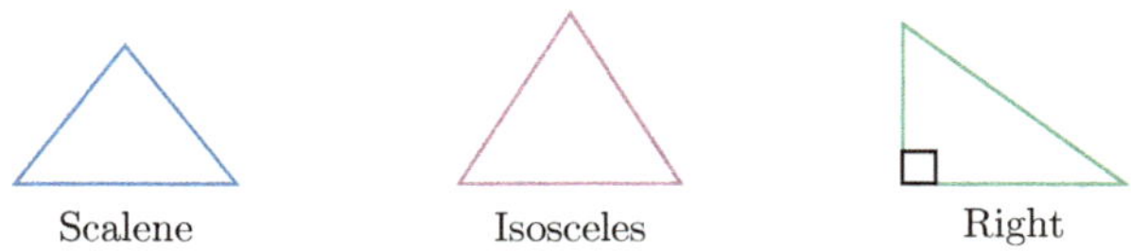

Quadrilaterals

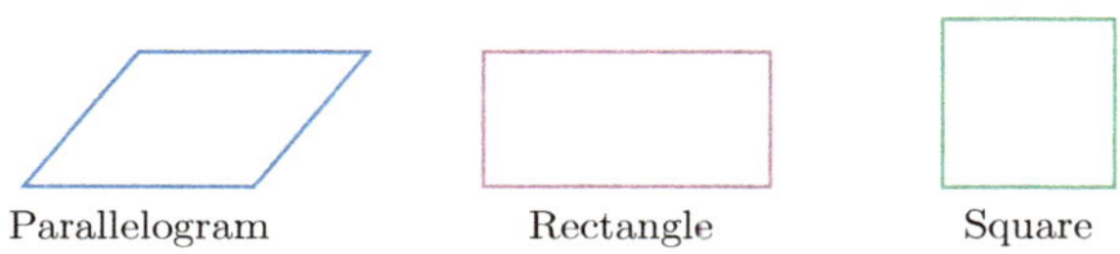

Circles

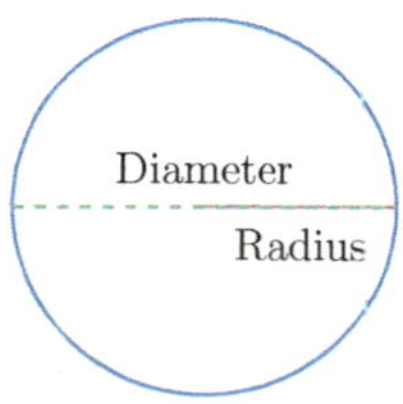

Area, Perimeter, Volume

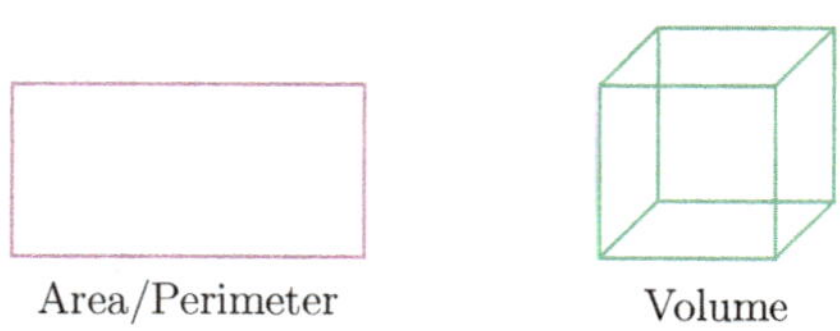

Transformations

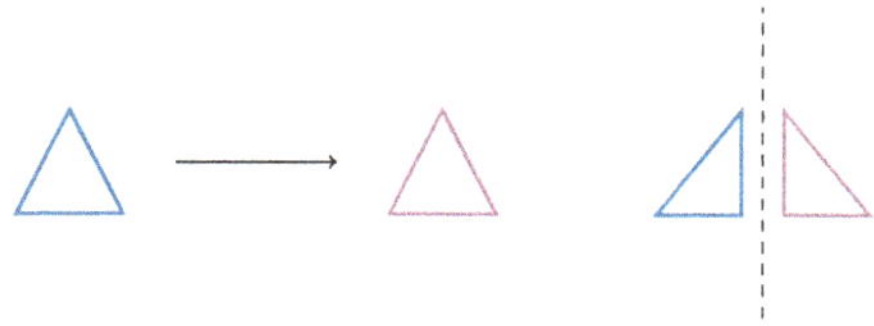

Coordinate Geometry

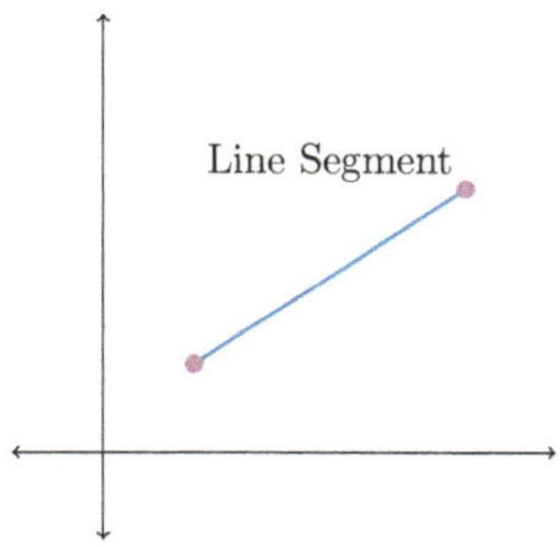

Similarity and Congruence

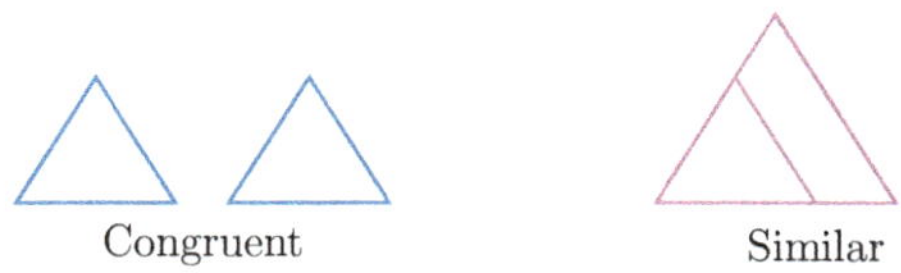

Proofs

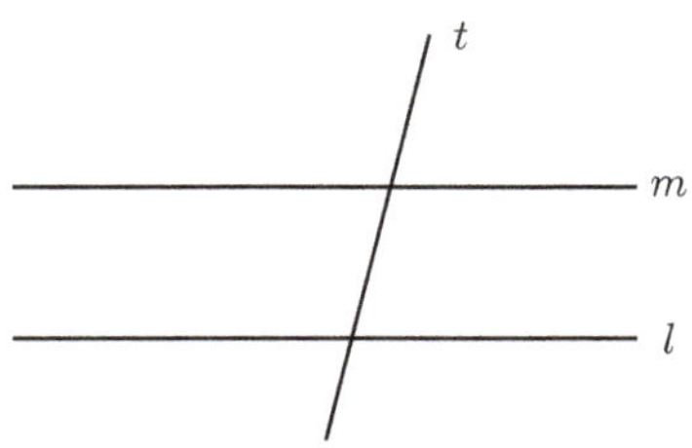

Proofs = angle logic

34.2 My Final Advice

- Draw pictures — always.
- Label everything — even your lunch.
- Check your work — twice.
- Don't panic — it's just shapes.
- And remember: Dad believes in you.

Memory Trick!

If you can think clearly, draw neatly, and stay patient, geometry becomes way less scary — promise.

If you made it this far, I'm proud of you. Geometry isn't always easy, but you stuck with it. And hey — we even had some laughs along the way.

Just remember:

You're smart. You're capable. And even on the days when you forget what a right angle is... God still loves you, stays with you, and isn't going anywhere.

Now go do your homework. And maybe clean your room while you're at it.

www.ingramcontent.com/pod-product-compliance
Lightning Source LLC
LaVergne TN
LVHW022106200926
848481LV00003B/1077